Fuzzy Logic: A Practical Approach

Fuzzy Logic: A Practical Approach

Jacob Lopez

Clanrye International
www.clanryeinternational.com

Clanrye International,
750 Third Avenue, 9th Floor,
New York, NY 10017, USA

ISBN: 978-1-64726-096-5

Cataloging-in-Publication Data

Fuzzy logic : a practical approach / Jacob Lopez.
p. cm.
Includes bibliographical references and index.
ISBN: 978-1-64726-096-5
1. Fuzzy logic. 2. Fuzzy systems. 3. Logic, Symbolic and mathematical.
4. Fuzzy mathematics. I. Lopez, Jacob.
QA9.64 .F89 2022
511.313--dc23

For information on all Clanrye International publications
visit our website at www.clanryeinternational.com

CLANRYE INTERNATIONAL

TABLE OF CONTENTS

PREFACE

This book has been written, keeping in view that students want more practical information. Thus, my aim has been to make it as comprehensive as possible for the readers. I would like to extend my thanks to my family and co-workers for their knowledge, support and encouragement all along.

Fuzzy logic is a type of many-valued logic which is used to handle the concept of partial truth. The truth value within this field can range from completely true to completely false. In fuzzy logic, the truth values of variables can be any real number between 0 and 1, where both extremes are inclusive. It depends on the observation that people make decisions on the basis of constant and non-numerical information. Fuzzy models are mathematical means for recognizing, representing, interpreting, manipulating and utilizing data and information that are vague and uncertain. It is applied in many fields such as control theory and artificial intelligence. The topics covered in this extensive book deal with the core aspects of fuzzy logic. It covers in detail some existent theories and innovative concepts revolving around this field. Those in search of information to further their knowledge will be greatly assisted by this book.

A brief description of the chapters is provided below for further understanding:

Chapter - What is Fuzzy Logic?

Fuzzy logic is used for handling the aspect of partial truth, where the truth-values lie between 0 and 1 as completely true or false. The control system of fuzzy logic is referred to as fuzzy control system. This chapter has been carefully written to provide an easy understanding of fuzzy logic.

Chapter - Diverse Aspects of Fuzzy Logic.

Some of the fundamentals of fuzzy logic include fuzzy numbers, fuzzy arithmetic, fuzzy mathematics, fuzzy sets, fuzzy rule, fuzzy associative matrix, fuzzy finite element method, Mamdani fuzzy model, etc. This is an introductory chapter which will briefly introduce all these fundamentals related to fuzzy logic.

Chapter - Formal Systems of Fuzzy Logic

The formal system of fuzzy logic is categorized into propositional fuzzy logic and predicate fuzzy logic. It consists of T-norm fuzzy logic, Monoidal T-norm basic logic, basic fuzzy logic, Godel-Dummett logic and Lukasiewicz logic. The topics elaborated in this chapter will help in gaining a better perspective about these formal systems of fuzzy logic.

Chapter - Processes in Fuzzy Logic

Fuzzification, negation criterion, minimum criterion, maximum criterion, fuzzy If-Then rules and defuzzification are a few processes that fall under the domain of fuzzy logic. This chapter closely examines these processes used in fuzzy logic to provide an extensive understanding of the subject.

Chapter - Applications

Fuzzy logic has been used in numerous applications such as facial pattern recognition, air conditioners, washing machines, vacuum cleaners, antiskid braking systems, transmission systems, control of subway systems and unmanned helicopters, robot navigation, etc. All these applications of fuzzy logic have been carefully analyzed in this chapter.

Jacob Lopez

What is Fuzzy Logic?

CHAPTER 1

Fuzzy logic is used for handling the aspect of partial truth, where the truth-values lie between 0 and 1 as completely true or false. The control system of fuzzy logic is referred to as fuzzy control system. This chapter has been carefully written to provide an easy understanding of fuzzy logic.

Fuzzy logic is an approach to computing based on "degrees of truth" rather than the usual "true or false" (1 or 0) Boolean logic on which the modern computer is based.

The idea of fuzzy logic was first advanced by Dr. Lotfi Zadeh of the University of California at Berkeley in the 1960s. Dr. Zadeh was working on the problem of computer understanding of natural language. Natural language (like most other activities in life and indeed the universe) is not easily translated into the absolute terms of 0 and 1. It may help to see fuzzy logic as the way reasoning really works and binary or Boolean logic is simply a special case of it.

Fuzzy logic includes 0 and 1 as extreme cases of truth (or "the state of matters" or "fact") but also includes the various states of truth in between so that, for example, the result of a comparison between two things could be not "tall" or "short" but ".38 of tallness."

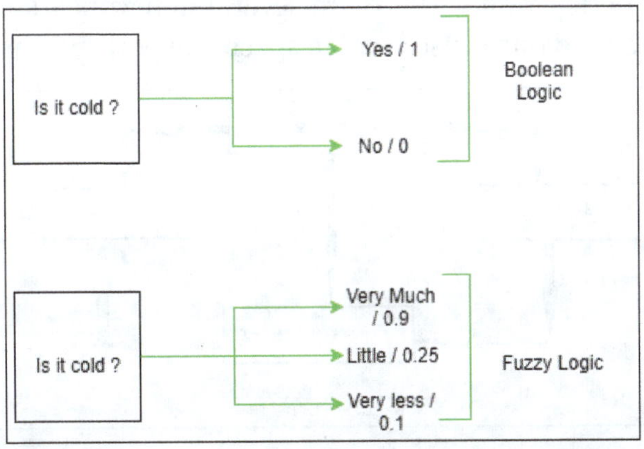

Fuzzy logic seems closer to the way our brains work. We aggregate data and form a number of partial truths which we aggregate further into higher truths which in turn, when certain thresholds are exceeded, cause certain further results such as motor reaction. A similar kind of process is used in neural networks, expert systems and other artificial intelligence applications. Fuzzy logic is essential to the development of human-like capabilities for AI, sometimes referred to as artificial general intelligence: the representation of generalized human cognitive abilities in software so that, faced with an unfamiliar task, the AI system could find a solution.

In boolean system truth value, 1.0 represents absolute truth value and 0.0 represents absolute false value. But in the fuzzy system, there is no logic for absolute truth and absolute false value. But in fuzzy logic, there is intermediate value too present which is partially true and partially false.

Architecture

Its Architecture contains four parts:

- RULE BASE: It contains the set of rules and the IF-THEN conditions provided by the experts to govern the decision making system, on the basis of linguistic information. Recent developments in fuzzy theory offer several effective methods for the design and tuning of fuzzy controllers. Most of these developments reduce the number of fuzzy rules.
- FUZZIFICATION: It is used to convert inputs i.e. crisp numbers into fuzzy sets. Crisp inputs are basically the exact inputs measured by sensors and passed into the control system for processing, such as temperature, pressure, rpm's, etc.
- INFERENCE ENGINE: It determines the matching degree of the current fuzzy input with respect to each rule and decides which rules are to be fired according to the input field. Next, the fired rules are combined to form the control actions.
- DEFUZZIFICATION: It is used to convert the fuzzy sets obtained by inference engine into a crisp value. There are several defuzzification methods available and the best suited one is used with a specific expert system to reduce the error.

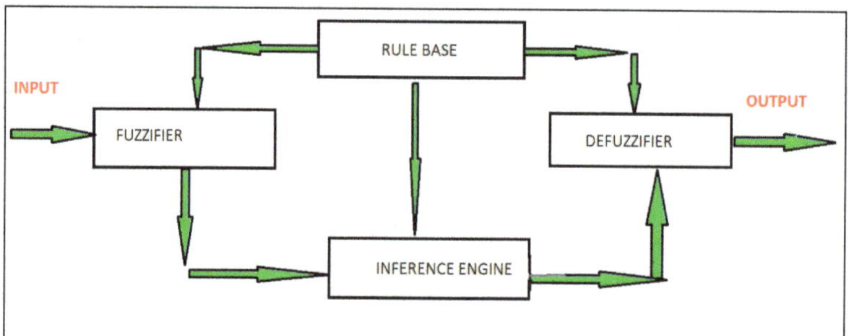

FUZZY LOGIC ARCHITECTURE

Membership Function

A graph that defines how each point in the input space is mapped to membership value between 0 and 1. Input space is often referred as the universe of discourse or universal set (u), which contain all the possible elements of concern in each particular application.

There are largely three types of fuzzifiers:

- Singleton fuzzifier.
- Gaussian fuzzifier.
- Trapezoidal or triangular fuzzifier.

Fuzzy Control

- It is a technique to embody human-like thinkings into a control system.
- It may not be designed to give accurate reasoning but it is designed to give acceptable reasoning.
- It can emulate human deductive thinking, that is, the process people use to infer conclusions from what they know.
- Any uncertainties can be easily dealt with the help of fuzzy logic.

Two Approaches

There are two basic approaches to fuzzy logic. The first one, proposed by P. Hajek and followed by Di Nola, Esteva, Gottwald, Godo, Montagna, Mundici and by a large series of students, is very close to the tradition of multi-valued logic. Indeed the deduction apparatus works on a set of hypotheses to give the corresponding set of logical consequences. This is obtained, as it is usual in multi-valued logic, once a set of designed truth values is fixed. We call, ungraded approach such a way to face fuzzy logic. Another approach was proposed by J. A. Goguen, J. Pavelka, V. Novak, G. Gerla and further authors and it is rather out of line with the tradition of multi-valued logic. Indeed, the deduction apparatus works on a given fuzzy subset of hypotheses to give the related fuzzy subset of logical consequences. We call graded approach such a way to face fuzzy logic.

The Ungraded Approach

In the ungraded approach a subset Des of [0, 1] is fixed whose elements are called designed truth degrees. The interpretation is that in Des there are the truth degrees which one considers sufficient to claim the validity of a formula. Usually one sets Des = {1}.

Definition: Let $([0, 1], \odot, \rightarrow, 0, 1)$ be a fixed standard algebra, and α be a formula. Then we say that a fuzzy interpretation (D, I) satisfies α provided that $\mathrm{Val}(I, \alpha)$ is a designed value. Let T be a theory, then (D, I) is a model of T if every formula in T is satisfied in (D, I). We write $T \models_{\odot} \alpha$ if every model of T satisfies α.

The deduction apparatus in the ungraded approach is defined by adopting the same paradigm of classical logic, i.e. a deduction relation $\models$ is defined by a suitable set of logical axioms and suitable inference rules. The fuzzy logic defined by ⊙ is axiomatizable provided that a deduction apparatus exists such that $\models$ coincides with $\models_{\odot}$. Unfortunately, the main fuzzy logics are not axiomatizable.

Theorem: In all the main fuzzy logics (in particular in Łukasievicz logic) the entailment relation $\models_{\odot}$ is not compact. This entails that these logics are not axiomatizable.

As an attempt to bypass such an obstacle, in the ungraded approach one proposes a different entailment relation related with the variety generated by a given triangular norm.

Definition: Given a standard algebra $([0,\ 1],\ \odot,\rightarrow,0,1)$, denote by $\mathrm{Varl}(\odot)$ the class of all linearly ordered algebras in the variety generated by ([0, 1], ⊙, →, 0, 1). Then a Varl(⊙)*-interpretation* is an interpretation in a valuation algebra belonging to Varl(⊙). Given a set T of formulas and a formula α, we write T $\models_{\mathrm{Varl}(\odot)}$ α provided that every safe Varl(⊙)-model of T is a safe Varl(⊙)-model of α.

In such a case, the resulting logic works well. In fact, the following theorem holds true.

Theorem: In all the main fuzzy logics (in particular in Łukasievicz logic) the entailment relation $\models_{\mathrm{Varl}(\odot)}$ is compact. This is in accordance with the fact that these logics are axiomatizable (provided that they are defined by referring to this relation).

Criticisms: A criticism for the ungraded approach, philosophical in nature, concerns its adequateness to represent the daily reasonings in which vague predicates occur. Moreover the structures in Varl(⊙) look rather unnatural. For example, in Varl(⊙) there are structures with infinitesimal truth values. Another criticism is that, while the completeness of [0, 1] assures that all the formulas are valuated, in the case we refer to the variety Varl(⊙), we are forced to admit interpretations for which there are unvaluated formulas.

The Graded Approach: Approximate Reasonings

The aim of any logic is to elaborate (uncomplete) information to obtain more explicit information. Now, in the case of fuzzy logic it is natural to admit an information like *"the truth values of α is between λ and μ"*, i.e. a constraint on the possible truth value of a formula. Now, observe that if we admit the usual interpretation of the negation, then $Val(I,\alpha)\leq\lambda$ if and only if $Val(I,\neg\alpha)\geq 1\text{-}\mu$. Then we can reduce all the interval constraints to lower bound constraints. In accordance in the graded approach one proposes the following definition.

Definition: Consider a fuzzy theory s, i.e. a fuzzy subset of formulas. Then a fuzzy interpretation (*D, I*) is a *model of* s, in brief $(D,I) \models s$ if $Val(I,\alpha) \geq s(\alpha)$. The *logical consequence operator* is the map $Lc:[0,\ 1]^F \rightarrow [0,\ 1]^F$ defined by setting:

$$\mathrm{Lc(s)(\alpha) = Inf\{Val(I,\alpha) : (D, I) \models s\}}.$$

Equivalently, we can refer to a graded entailment relation λ by writng s $\models_{\lambda}$ α where $\lambda = \mathrm{Inf\{Val(I,\alpha) : (D,I) \models s\}}$.

These definitions are in accordance with the fact that s represents a system of "lower bound constraints" on the unknown truth value of the formulas. Moreover, Lc(s) is the better lower bound constraint we can find given s. In the graded approach we can obtain a deduction apparatus by extending Hilbert's approach as follows.

Definition: A fuzzy inference rule is a pair r = (syn,sem) where syn, the syntactical part, is a partial n-ary operation in F (i.e. an inference rule in the usual sense) and sem, the semantic part, is an n-ary join-preserving operation in [0,1]. An evaluated syntax is a structure (la, R) where la is a fuzzy set of formulas we call fuzzy subset of logical axioms, and R is a set of fuzzy inference rules.

Usually, n = 2 and $sem(\lambda_1,\lambda_2)$ is a product like $\lambda_1 \odot \lambda_2$. As an example, the fuzzy Modus Ponens is defined by assuming that the domain of syn is the set {(α, α→β): α,β are in F}, by setting syn(α, α→β) = β and by assuming that $sem(\lambda, \mu) = \lambda \odot \mu$. This rule says that:

- If we are able to prove α at degree λ,
- And α → β at degree μ,
- Then we can prove β at degree λ⊙μ.

Definition: A proof π of a formula α is a sequence $\alpha_1,\ldots,\alpha_m$ of formulas such that $\alpha_m = \alpha$, together with a sequence of related justifications. This means that, for every formula α_i, we have to specify whether:

- α_i is assumed as a logical axiom or.
- α_i is assumed as an hypothesis or.
- α_i is obtained by a rule (in this case we have to indicate the rule and the formulas from $\alpha_1,\ldots,\alpha_{i-1}$ used to obtain α_i).

The justifications are necessary to valuate the proofs. Indeed, let s be the fuzzy subset of proper axioms and, for every i ≤ m denote by π(i) the proof $\alpha_1,\ldots,\alpha_i$. Then the information furnished by π given s is the value Val(π, s) is defined by induction on m by setting:

$Val(\pi,s) = la(\alpha_m)$ if α_m is assumed as a logical axiom.

$Val(\pi,s) = s(\alpha_m)$ if α_m is assumed as an hypothesis.

$Val(\pi,s) = sem(Val(\pi(\mathrm{i}_1),s),\ldots,Val(\pi(i_n),s))$ if there is a fuzzy rule (*syn*,*sem*) such that $\alpha_m = syn(\alpha_{i1},\ldots,\alpha_{in})$ with $i_1 < m,\ldots,i_n < m$.

Now, unlike the usual deduction systems, in a fuzzy deduction system, different proofs of a same formula α may give different pieces of information on the truth degree of α. Then, we have to *"fuse"* all these informations.

Definition: The *deduction operator* is the operator D defined by setting:

$D(s)\ (\alpha) = Sup\{Val(\pi,s)|\ \pi$ is a proof of $\alpha\}$.

A fuzzy logic is *axiomatizable* if there is a fuzzy deduction system such that $Lc = D$.

In Novak 2007 one proves the following basic result.

Proposition: In the graded approach Łukasiewicz first order logic is axiomatizable.

Notice also that, under some natural hypotheses, a fuzzy propositional logic is axiomatizable if and only if the logical connectives are interpreted by continuous functions.

Criticisms: A criticism for such an approach is that is not sufficiently flexible. As an example let (D, I) be a fuzzy model of a fuzzy theory s. It is evident that both the assignments (D, I) and s cannot be considered definitive and precise since both depends on a subjective valuation. Assume that either (D, I) or s is subject to a slightly variation as a consequence of a tuning process, an essential component in all the applications in fuzzy mathematics. Then it is possible that (D, I) ceases completely to be a model of s while it should be natural to expect (D, I) is again a model of s at some degree.

Continuous Logic

A very important precursor of fuzzy logic is the Continuous logic proposed by Chang and Keisler since 1966. In their book all the model theoretical notions of classical logic are extended to first order multi-valued logic. As an example, the notions of quotient, direct product, ultraproduct are defined and examined. Recently, an interesting reformulation of continuous logic was proposed to give a basis for a model theory for the various kinds of "metric" structures arising in functional analysis and probability theory. Such a reformulation is obtained by referring to the interval [0, 1] and by interpreting the logical connectives by a class of continuous functions able to define a complete approximation system. Also one considers models in which a pseudo-metric is defined to interpret the symbol =. One assumes that all the predicates and function symbols are uniformly continuous with respect such a pseudo-metric. As an example, the finitely additive probabilities can be considered models of a continuous logic once we consider a language for Boolean algebras equipped with a vague monadic predicate p such that the interpretation of p(x) is "the event x is probable".

Fuzzy Logic with no Truth-functional Semantics

Fuzzy logic extends beyond the truth-functional tradiction of multi-valued logic. The following are two examples:

Necessity Logic

Assume that the deduction apparatus of classical first order logic is presented by a suitable set la of logical axioms, by the MP-rule and the Generalization rule. Then a fuzzy deduction system is obtained by considering as a fuzzy subset of logical axioms the characteristic function of la and as fuzzy inference rules the fuzzy Modus Ponens and the extension of the Generalization Rule obtained by assuming that if we prove α at degree λ then we obtain ∀ xα(x) at the same degree λ. In suh a case it is easy to see that if s is a fuzzy theory, then:

D(s)(α) = 1 i α is a logically true formula,

$D(s)(\alpha)$ =Sup$\{s(\alpha_1)\odot ...\odot s(\alpha_n) : \alpha_1,..., \alpha_n \models \alpha\}$, otherwise.

By recalling that the existential quantifier is interpreted by the supremum operator, such a formula arises from a multivalued valuation of the (metalogical) claim: "α is a consequence of the fuzzy subset s *of axioms provided there are formulas* α_1, ...,α_n in s *able to prove* α". In such a case the vagueness originates from s, i.e., from the notion of "hypothesis". Moreover s(α) is not a truth degree but rather a degree of "preference" or "acceptability" for α. For example, let ⊙ be the minimum, T be a system of axioms for set theory such that the choice axiom CA does not depend on T. Then we can consider the fuzzy subset of axioms s defined by setting:

s(α) = 1 if α ϵ T,

s(α) = 0.8 if α = CA ,

s(α) = 0 otherwise.

A simple calculation shows that:

D(s)(α) = 1 if α is a theorem of T,

D(s)(α) = 0.8 if we cannot prove α from T but α is a theorem of T + CA,

D(s)(α) = 0 otherwise .

Then, despite the fact that no vague predicate is considered in set theory, in the metalanguage we can consider a vague meta-predicate as "is acceptable" and to represent it by a suitable fuzzy subset s.

Similarity Logic

In accordance with the ideas of M. S. Ying we can extend necessity logic by introducing a similarity relation among the predicates. As an example, consider an inference like:

- Since x is a thriller ⇛ x good for me +.
- and b is a detective story +.
- and "detective story" is synonymous of "thriller".
- then "b is good for me".

Now the synonymy is a vague notion we can represent by a suitable similarity in the set W of English worlds, i.e. a fuzzy relation e such that:

- $e(x, x) = 1$ (reflexivity),
- $e(x, z) \odot e(z, y) \leq e(x, y)$ (transitivity),
- $e(x, y) = e(y, x)$ (symmetry).

Also, as it is usual in fuzzy logic, it is natural to admit that the truth degree of the conclusion "b is good for me" depends on the degree of similarity between the predicates "detective story" and "thriller", obviously. The structure of the corresponding fuzzy inference rule is:

If α was proven at degree λ,

and α'→ β at degree μ,

then β is proven at degree $\lambda \odot \mu \odot e(\alpha, \alpha')$.

Every inference rule can be extended in a similar way, i.e. by relaxing the precise matching of the identity with the approximate matching of a similarity. These ideas are also on the basis for a similarity-based fuzzy logic programming.

Effectiveness

A test to analyze the effectiveness in the ungraded approach to fuzzy logic is to refer to the set of tautologies. Now, since two entailment relations are defined, we have to consider two corresponding notions of tautology.

Definition: Given a standard algebra $([0, 1], \odot, \rightarrow, 0, 1)$ a formula α is a standard tautology if it is satisfied in every fuzzy interpretation in $([0,1], \odot, \rightarrow, 0, 1)$. The formula α is a general tautology if it is satisfied in every safe Varl(⊙)-interpretation.

In the first case the following negative result holds true.

Theorem: In the case of Łukasiewicz and product logic the set of standard tautologies is not recursively enumerable.

Such a fact gives a further confirm on the impossibility of an axiomatization of the entailment relation $\models$ and it leads to focalize the attention on $\models_{Varl(\odot)}$. At this regard one proves the following theorem:

Theorem: For each continuous t-norm ⊙, the set of general ⊙-tautologies in first order logic is Σ_1-complete (and therefore recursively enumerable).

In the case of the graded approach to face the question of the effectiveness we have to refer to the notion of effectiveness for fuzzy subset theory. A first proposal in such a direction was made by E.S. Santos by the notions of fuzzy Turing machine. Successively, in Biacino and Gerla the following definition was proposed where $\ddot{U}$ denotes the set of rational numbers in [0, 1].

Definition: A fuzzy subset $s : S \rightarrow [0,1]$ of a set S is recursively enumerable if a recursive map $h : S \times N \rightarrow \ddot{U}$ exists such that, for every x in S, the function h(x, n) is increasing with respect to n and s(x) = lim h(x,n). We say that s is decidable if both s and its complement –s are recursively enumerable.

An extension of such a theory to the general case of the L-subsets is proposed in Gerla where one refers to the theory of effective domains. It is an open question to give supports for a Church thesis for fuzzy set theory claiming that the proposed notion of recursive enumerability for fuzzy subsets is the adequate one.

In Gerla one proves the following theorem where we refer to fuzzy logics whose deduction apparatus satisfies some obvious effectiveness properties.

Theorem: Given an axiomatizable fuzzy logic, its fuzzy subset $D(\emptyset)$ of tautologies is recursively enumerable. In particular the fuzzy subset of tautologies in Łukasiewicz logic is recursively enumerable in spite of the non recursive enumerability of its cut $\{\alpha : D(\emptyset)(\alpha) = 1\}$.

Advantages of Fuzzy Logic Systems

- This system can work with any type of inputs whether it is imprecise, distorted or noisy input information.
- The construction of Fuzzy Logic Systems is easy and understandable.
- Fuzzy logic comes with mathematical concepts of set theory and the reasoning of that is quite simple.
- It provides a very efficient solution to complex problems in all fields of life as it resembles human reasoning and decision making.
- The algorithms can be described with little data, so little memory is required.

Disadvantages of Fuzzy Logic Systems

- Many researchers proposed different ways to solve a given problem through fuzzy logic which lead to ambiguity. There is no systematic approach to solve a given problem through fuzzy logic.
- Proof of its characteristics is difficult or impossible in most cases because every time we do not get mathematical description of our approach.
- As fuzzy logic works on precise as well as imprecise data so most of the time accuracy is compromised.

Application

- It is used in the aerospace field for altitude control of spacecraft and satellite.
- It has used in the automotive system for speed control, traffic control.
- It is used for decision making support systems and personal evaluation in the large company business.
- It has application in chemical industry for controlling the pH, drying, chemical distillation process.
- Fuzzy logic are used in Natural language processing and various intensive applications in Artificial Intelligence.
- Fuzzy logic are extensively used in modern control systems such as expert systems.
- Fuzzy Logic is used with Neural Networks as it mimics how a person would make decisions, only much faster. It is done by Aggregation of data and changing into more meaningful data by forming partial truths as Fuzzy sets.

FUZZY LOGIC CONTROL

Fuzzy logic control is a heuristic approach that easily embeds the knowledge and key elements of human thinking in the design of nonlinear controllers. Qualitative and heuristic considerations, which cannot be handled by conventional control theory, can be used for control purposes in a systematic form, applying fuzzy control concepts. Fuzzy logic control does not need an accurate mathematical model, can work with imprecise inputs, can handle nonlinearity, and can present disturbance insensitivity greater than the most nonlinear controllers. Fuzzy logic controllers usually outperform other controllers in complex, nonlinear, or undefined systems for which a good practical knowledge exists.

Fuzzy logic controllers are based on fuzzy sets, that is, classes of objects in which the transition from membership to nonmembership is smooth rather than abrupt. Therefore, boundaries of fuzzy sets can be vague and ambiguous, making them useful for approximation models.

The first step in the fuzzy controller synthesis procedure is to define the input and output variables of the fuzzy controller. This is done accordingly with the expected function of the controller. There are no general rules to select those variables, although typically the variables chosen are the states of the controlled system, their errors, error variation, and/or error accumulation. In switching power converters, the fuzzy controller input variables are commonly the output voltage or current error and/or the variation or accumulation of this error. The output variables u(k) of the fuzzy controller can define the converter duty cycle or a reference current to be applied in an inner current-mode PI or a sliding-mode controller.

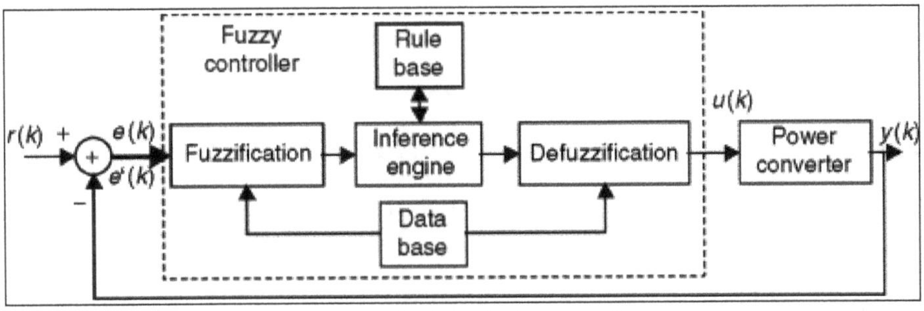

Structure of a fuzzy logic controller.

The fuzzy controller rules are usually formulated in linguistic terms. Thus, the use of linguistic variables and fuzzy sets implies the fuzzification procedure, that is, the mapping of the input variables into suitable linguistics values.

Rule evaluation or decision-making infers, using an inference engine, the fuzzy control action from the knowledge of the fuzzy rules and the linguistic variable definition.

The output of a fuzzy controller is a fuzzy set, and thus, it is necessary to perform a defuzzification procedure, that is, the conversion of the inferred fuzzy result to a nonfuzzy (crisp) control action, that better represents the fuzzy one. This last step obtains the crisp value for the controller output u(k).

These steps can be implemented online or offline. Online implementation, useful if an adaptive controller is intended, performs real-time inference to obtain the controller output and needs a fast enough processor. Off-line implementation employs a lookup table built according to the set of all possible combinations of input variables. To obtain this lookup table, the input values in a quantified range are converted (fuzzification) into fuzzy variables (linguistic). The fuzzy set output, obtained by the inference or decision-making engine according to linguistic control rules (designed by the knowledge expert), is then, converted into numeric controller output values (defuzzification). The table contains the output for all the combinations of quantified input entries. Off-line process can actually reduce the controller actuation time since the only effort is limited to consulting the table at each iteration.

Fuzzy Logic in Control Systems

A control system is an arrangement of physical components designed to alter another physical system so that this system exhibits certain desired characteristics. Following are some reasons of using Fuzzy Logic in Control Systems:

- While applying traditional control, one needs to know about the model and the objective function formulated in precise terms. This makes it very difficult to apply in many cases.

- By applying fuzzy logic for control we can utilize the human expertise and experience for designing a controller.
- The fuzzy control rules, basically the IF-THEN rules, can be best utilized in designing a controller.

Assumptions in Fuzzy Logic Control (FLC) Design

While designing fuzzy control system, the following six basic assumptions should be made:

- The plant is observable and controllable: It must be assumed that the input, output as well as state variables are available for observation and controlling purpose.
- Existence of a knowledge body: It must be assumed that there exist a knowledge body having linguistic rules and a set of input-output data set from which rules can be extracted.
- Existence of solution: It must be assumed that there exists a solution.
- 'Good enough' solution is enough: The control engineering must look for 'good enough' solution rather than an optimum one.
- Range of precision: Fuzzy logic controller must be designed within an acceptable range of precision.
- Issues regarding stability and optimality: The issues of stability and optimality must be open in designing Fuzzy logic controller rather than addressed explicitly.

Architecture of Fuzzy Logic Control

The following diagram shows the architecture of Fuzzy Logic Control (FLC).

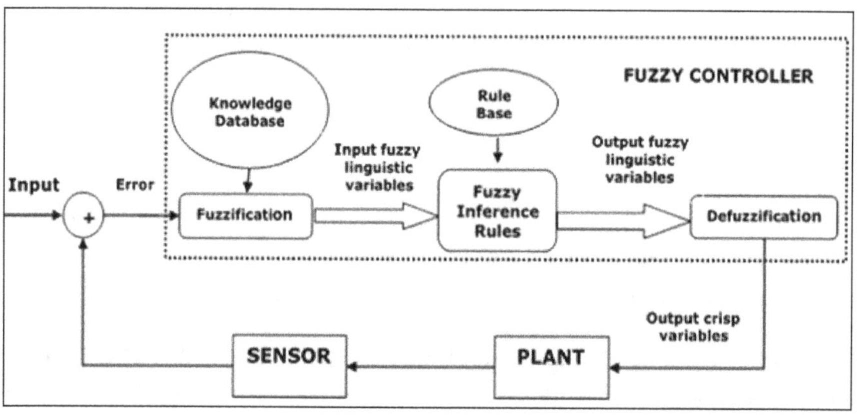

Major Components of FLC

Following are the major components of the FLC:

- Fuzzifier: The role of fuzzifier is to convert the crisp input values into fuzzy values.
- Fuzzy Knowledge Base: It stores the knowledge about all the input-output fuzzy relationships. It also has the membership function which defines the input variables to the fuzzy rule base and the output variables to the plant under control.
- Fuzzy Rule Base: It stores the knowledge about the operation of the process of domain.
- Inference Engine: It acts as a kernel of any FLC. Basically it simulates human decisions by performing approximate reasoning.
- Defuzzifier: The role of defuzzifier is to convert the fuzzy values into crisp values getting from fuzzy inference engine.

Steps in Designing FLC

Following are the steps involved in designing FLC:

- Identification of variables: Here, the input, output and state variables must be identified of the plant which is under consideration.
- Fuzzy subset configuration: The universe of information is divided into number of fuzzy subsets and each subset is assigned a linguistic label. Always make sure that these fuzzy subsets include all the elements of universe.
- Obtaining membership function: Now obtain the membership function for each fuzzy subset that we get in the above step.
- Fuzzy rule base configuration: Now formulate the fuzzy rule base by assigning relationship between fuzzy input and output.
- Fuzzification: The fuzzification process is initiated in this step.
- Combining fuzzy outputs: By applying fuzzy approximate reasoning, locate the fuzzy output and merge them.
- Defuzzification: Finally, initiate defuzzification process to form a crisp output.

Advantages of Fuzzy Logic Control

- Cheaper: Developing a FLC is comparatively cheaper than developing model based or other controller in terms of performance.

- Robust: FLCs are more robust than PID controllers because of their capability to cover a huge range of operating conditions.
- Customizable: FLCs are customizable.
- Emulate human deductive thinking: Basically FLC is designed to emulate human deductive thinking, the process people use to infer conclusion from what they know.
- Reliability: FLC is more reliable than conventional control system.
- Efficiency: Fuzzy logic provides more efficiency when applied in control system.

Disadvantages of Fuzzy Logic Control

- Requires lots of data: FLC needs lots of data to be applied.
- Useful in case of moderate historical data: FLC is not useful for programs much smaller or larger than historical data.
- Needs high human expertise: This is one drawback as the accuracy of the system depends on the knowledge and expertise of human beings.
- Needs regular updating of rules: The rules must be updated with time.

References

- Fuzzy-logic: searchenterpriseai.techtarget.com, Retrieved 14 July, 2019
- Fuzzy-logic-introduction: geeksforgeeks.org, Retrieved 04 March, 2019
- Formal-fuzzy-logic: en.citizendium.org, Retrieved 25 August, 2019
- Fuzzy-logic-control, engineering: sciencedirect.com, Retrieved 28 May, 2019
- Fuzzy-logic-control-system, fuzzy-logic: tutorialspoint.com, Retrieved 19 January, 2019

Diverse Aspects of Fuzzy Logic

CHAPTER 2

Some of the fundamentals of fuzzy logic include fuzzy numbers, fuzzy arithmetic, fuzzy mathematics, fuzzy sets, fuzzy rule, fuzzy associative matrix, fuzzy finite element method, Mamdani fuzzy model, etc. This is an introductory chapter which will briefly introduce all these fundamentals related to fuzzy logic.

FUZZY NUMBERS

A fuzzy number is a quantity whose value is imprecise, rather than exact as is the case with "ordinary" (single-valued) numbers. Any fuzzy number can be thought of as a function whose domain is a specified set (usually the set of real numbers, and whose range is the span of non-negative real numbers between, and including, 0 and 1000. Each numerical value in the domain is assigned a specific "grade of membership" where 0 represents the smallest possible grade, and 1000 is the largest possible grade.

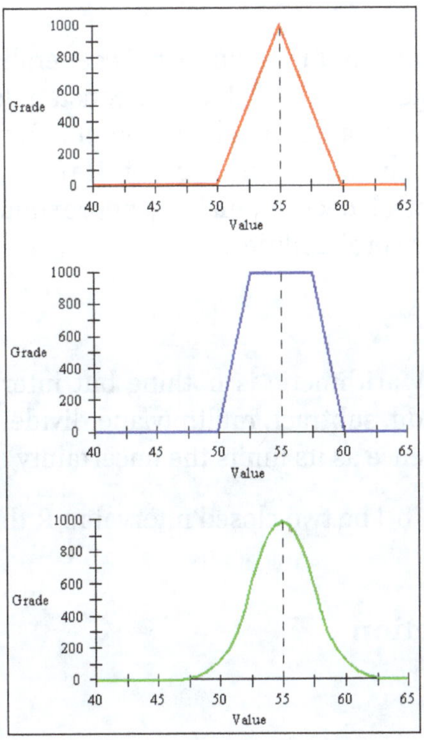

In many respects, fuzzy numbers depict the physical world more realistically than single-valued numbers. Suppose, for example, that you are driving along a highway where the speed limit is 55 miles an hour (mph). You try to hold your speed at exactly 55 mph, but your car lacks "cruise control," so your speed varies from moment to moment. If you graph your instantaneous speed over a period of several minutes and then plot the result in rectangular coordinates, you will get a function that looks like one of the curves.

The red curve (top) represents a triangular fuzzy number; the blue curve(middle) shows a trapezoidal fuzzy number; the green curve (bottom) illustrates a bell-shaped fuzzy number. These three functions, known as membership functions , are all convex (the grade starts at zero, rises to a maximum, and then declines to zero again as the domain increases). However, some fuzzy numbers have concave, irregular, or even chaotic membership functions. There is no restriction on the shape of the membership curve, as long as each value in the domain corresponds to one and only one grade in the range, and the grade is never less than 0 nor more than 1000.

Fuzzy numbers are used in statistics, computer programming, engineering(especially communications), and experimental science. The concept takes into account the fact that all phenomena in the physical universe have a degree of inherent uncertainty.

FUZZY ARITHMETIC

Fuzzy arithmetic or arithmetic of fuzzy numbers is generalisation of interval arithmetic, where rather than considering intervals at one constant level only, several levels are considered in [0, 1]. This is because, the definition of a fuzzy set which allows degree of membership for an element of the universal set. It plays an important role in many applications, say, fuzzy control, decision making, approximate reasoning, optimization and statistics with imprecise probabilities.

Interval Arithmetic

The fundamentals of fuzzy arithmetic is nothing but interval arithmetic. In a given closed interval R, how to add, subtract, multiply and divide. These interval in R is also called an interval of confidence as its limits the uncertainty of data to an interval.

Let $A = [a_1, a_2]$ and $B = [b_1, b_2]$ be two closed interval in R then we have following definitions:

Addition and Subtraction

If $x \in [a_1, a_2]$, $y \in [b_1, b_2]$,

then $x + y \in [a_1 + b_1, a_2 + b_2]$ and $x - y \in [a_1 - b_1, a_2 - b_2]$

therefore,

$$A + B = [a_1, a_2] + [b_1, b_2] = [a_1 + b_1, a_2 + b_2].$$
$$A - B = [a_1, a_2] - [b_1, b_2] = [a_1 - b_2, a_2 - b_1].$$

Example:

$$[2,5] + [1,3] = [3,8].$$
$$[0,1] - [-6,5] = [-5,7].$$

Image of an Interval

If $x \in [a_1, a_2]$ then its image $-x \in -[-a_2, -a_1]$. Therefore the image of A is denoted by $\overline{A}$ is defined as:

$$\overline{A} = [a_1, a_2] = [-a_2, -a_1].$$

Multiplication

The multiplication of two closed intervals $A = [a_1, a_2]$ and $B = [b_1, b_2]$ of R denoted by A.B is defined as:

$$A \cdot B = [\min(a_1 b_1, a_1 b_2, a_2 b_1, a_2 b_2), \max(a_1 b_1, a_1 b_2, a_2 b_1, a_2 b_2)].$$

Example:

$$[-1,1] \cdot [-2,0.5] = [\min(2,0.5,-2,-0.5), \max(2,0.5,-2,-0.5)] = [-2,2].$$

Scalar Multiplication and Inverse

Let $A = [a_1, a_2]$ be a closed interval in R and $k \in R$ identifying the scalar k as the closed interval [k, k], the scalar multiplication k.A is defined as:

$$k.A = [k, k] \cdot [a_1, a_2] = [ka_1, ka_2].$$

For $A = [a_1, a_2] \in R$ if $x \in [a_1, a_2]$ and if $0 \notin [a_1, a_2]$ then $\frac{1}{x} \in \left[\frac{1}{a_2}, \frac{1}{a_1}\right]$.

Therefore the inverse of A is denoted by A^{-1} and it is defined as:

$$A^{-1} = [a_1, a_2]^{-1} = \left[\frac{1}{a_2}, \frac{1}{a_1}\right] \text{ provided } 0 \notin [a_1, a_2].$$

Division

The division of two closed intervals A = $[a_1, a_2]$ and B = $[b_1, b_2]$ of R denoted by A/B is defined as the multiplication $[a_1, a_2]$ and $\left[\frac{1}{b_2}\frac{1}{b_1}\right]$ provided $0 \notin [b_1, b_2]$.

Therefore A/B = $[a_1, a_2] / [b_1, b_2]$.

$$= [a_1, a_2]\left[\frac{1}{b_2}, \frac{1}{b_1}\right].$$

$$= \left[\min\left(\frac{a_1}{b_2}, \frac{a_1}{b_1}, \frac{a_2}{b_2}, \frac{a_2}{b_1}\right) \max\left(\frac{a_1}{b_2}, \frac{a_1}{b_1}, \frac{a_2}{b_2}, \frac{a_2}{b_1}\right)\right].$$

Example:

$[4, 10] / [1, 2] = [2, 10]$.

Arithmetic operations on closed intervals satisfy some useful properties, to overview, let A = $[a_1, a_2]$, B = $[b_1, b_2]$, C = $[c_1, c_2]$, 0 = [0, 0] and 1= [1, 1], using these symbols, the properties are formulated as follows:

- $A + B = B + A$ and $A \cdot B = B \cdot A$ (Commutative).
- $(A + B) + C = A + (B + C)$ and $A \cdot (B. C) = (A \cdot B) \cdot C$ (Associative).
- $A = 0 + A = A + 0$ and $A = 1 \cdot A = A \cdot 1$ (Identity).
- $A \cdot (B + C) \subseteq A \cdot B + A \cdot C$ (Subdistributivity).
- If $b \cdot c \geq 0$, forevery $b \in B$, $c \in C$, then $A \cdot (B + C) = A \cdot B + A \cdot C$(Distributive).

  Furthermore, if A = [a, a] then $a \cdot (B + C) = a \cdot B + a \cdot C$
- 0 in A – A and 1 in A/A.
- If $A \subseteq E$ and $B \subseteq F$ then $A + B \subseteq E + F$, $A - B \subseteq E - F$, $A \cdot B \subseteq E \cdot F$, $A/B \subseteq E/F$

  (Inclusion monotonicity).

As an example we prove only the less obvious properties of Subdistributivity. First we have:

$$A \cdot (B+C) = \{a \cdot (b + c) / a \in A, b \in B, c \in C\}.$$

$$= \{a \cdot b + a \cdot c / a \in A, b \in B, c \in C\}$$

$$\subseteq \{a \cdot b + a' \cdot c / a \cdot a' \in A, b \in B, c \in C\} = A \cdot B + A \cdot C.$$

Hence $A \cdot (B + C) \subseteq A \cdot B + A \cdot C$.

Assume now without loss of generality, that $b_1 \geq 0$, $c_1 \geq 0$. Then we have to consider the following three cases:

- If $a_1 \geq 0$ then $A \cdot (B + C) = [a_1 \cdot (b_1 + c_1), a_2 \cdot (b_2 + c_2)]$.

  $= [a_1 \cdot b_1 + a_2 \cdot b_2] + [a_1 \cdot c_1, a_2 \cdot c_2]$.

  $= A \cdot B + A \cdot C$.

- If $a_1 < 0$ and $a_2 \leq 0$ then $-a_2 \geq 0$, $(-A) = [-a_2, -a_1]$ and

  $(-A) \cdot (B + C) = (-A) \cdot B + (-A) \cdot C$.

  Hence $A \cdot (B + C) = A \cdot B + A \cdot C$.

- If $a_1 < 0$ and $a_2 > 0$ then $A \cdot (B + C) = [a_1 \cdot (b_2 + c_2), a_2 \cdot (b_2 + c_2)]$.

  $= [a_1 \cdot b_2, a_2 \cdot b_2] + [a_1 \cdot c_2, a_2 \cdot c_2]$.

  $= A \cdot B + A \cdot C$.

To show that distributivity does not hold in general,

Let $A = [0, 1]$, $B = [1, 2]$, $C = [-2, -1]$,

then $A \cdot B = [0, 2]$, $A \cdot C = [-2, 0]$, $B + C = [-1, 1]$ and

$A \cdot (B + C) = [-1, 1] \subset [2, 2] = A \cdot B + A \cdot C$.

Definition: (Max ∨ and Min ∧ Operations)

Let $A = [a_1, a_2]$ and $B = [b_1, b_2]$ be two closed intervals in R then the Max ∨ and min ∧ operations on A and B are defined as:

$$A \vee B = [a_1, a_2] \vee [b_1, b_2] = [a_1 \vee b_1, a_2 \vee b_2].$$
$$A \wedge B = [a_1, a_2] \wedge [b_1, b_2] = [a_1 \wedge b_1, a_2 \wedge b_2].$$

FUZZY MATHEMATICS

Fuzzy mathematics forms a branch of mathematics related to fuzzy set theory and fuzzy logic. It started in 1965 after the publication of Lotfi Asker Zadeh's seminal work Fuzzy sets. A fuzzy subset A of a set X is a function A:X→L, where L is the interval [0,1]. This function is also called a membership function. A membership function is a generalization of a characteristic function or an indicator function of a subset defined for L = {0,1}. More generally, one can use a complete lattice L in a definition of a fuzzy subset A .

The evolution of the fuzzification of mathematical concepts can be broken down into three stages:

- Straightforward fuzzification during the sixties and seventies.
- The explosion of the possible choices in the generalization process during the eighties.
- The standardization, axiomatization and L-fuzzification in the nineties.

Usually, a fuzzification of mathematical concepts is based on a generalization of these concepts from characteristic functions to membership functions. Let A and B be two fuzzy subsets of X. Intersection $A \cap B$ and union $A \cup B$ are defined as follows: $(A \cap B)(x) = \min(A(x),B(x))$, $(A \cup B)(x) = \max(A(x),B(x))$ for all $x \in X$. Instead of min and max one can use t-norm and t-conorm, respectively , for example, min(a, b) can be replaced by multiplication ab. A straightforward fuzzification is usually based on min and max operations because in this case more properties of traditional mathematics can be extended to the fuzzy case.

A very important generalization principle used in fuzzification of algebraic operations is a closure property. Let * be a binary operation on X. The closure property for a fuzzy subset A of X is that for all $x, y \in X$, $A(x*y) \geq \min(A(x),A(y))$. Let (G,*) be a group and A a fuzzy subset of G. Then A is a fuzzy subgroup of G if for all x, y in G, $A(x*y^{-1}) \geq \min(A(x),A(y^{-1}))$.

A similar generalization principle is used, for example, for fuzzification of the transitivity property. Let R be a fuzzy relation in X, i.e. R is a fuzzy subset of X×X. Then R is transitive if for all x, y, z in X, $R(x, z) \geq \min(R(x, y),R(y, z))$.

FUZZY SETS

Fuzzy sets can be considered as an extension and gross oversimplification of classical sets. It can be best understood in the context of set membership. Basically it allows partial membership which means that it contain elements that have varying degrees of membership in the set. From this, we can understand the difference between classical set and fuzzy set. Classical set contains elements that satisfy precise properties of membership while fuzzy set contains elements that satisfy imprecise properties of membership.

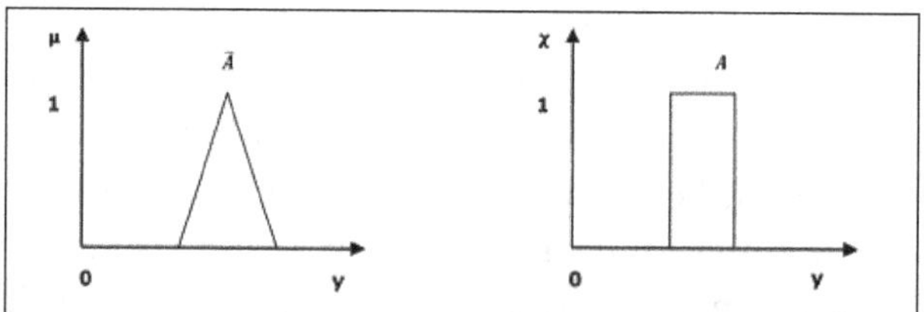

Membership Function of Fuzzy set Ã Membership Function of classical set A

Mathematical Concept

A fuzzy set $\tilde{A}$ in the universe of information U can be defined as a set of ordered pairs and it can be represented mathematically as:

$$\tilde{A} = \{(y, \mu_{\tilde{A}}(y)) | y \in U\}$$

Here $\mu_{\tilde{A}}(y) =$ degree of membership of y in {A}, assumes values in the range from 0 to 1, i.e., $\mu_{\tilde{A}}(y) \in [0,1]$.

Representation of Fuzzy Set

Let us now consider two cases of universe of information and understand how a fuzzy set can be represented.

Case 1

When universe of information U is discrete and finite:

$$\tilde{A} = \left\{ \frac{\mu_{\tilde{A}}(y_1)}{y_1} + \frac{\mu_{\tilde{A}}(y_2)}{y_2} + \frac{\mu_{\tilde{A}}(y_3)}{y_3} + \dots \right\}$$

$$= \left\{ \sum_{i=1}^{n} \frac{\mu_{\tilde{A}}(y_i)}{y_i} \right\}$$

Case 2

When universe of information U is continuous and infinite:

$$\tilde{A} = \left\{ \int \frac{\mu_{\tilde{A}}(y)}{y} \right\}$$

The summation symbol represents the collection of each element.

Operations on Fuzzy Sets

Having two fuzzy sets $\tilde{A}$ and $\tilde{B}$, the universe of information UU and an element y of the universe, the following relations express the union, intersection and complement operation on fuzzy sets.

Union/Fuzzy 'OR'

Let us consider the following representation to understand how the Union/Fuzzy 'OR' relation works:

$$\mu_{\tilde{A} \cup \tilde{B}}(y) = \mu_{\tilde{A}} \vee \mu_{\tilde{B}} \quad \forall y \in U$$

Here ∨ represents the 'max' operation.

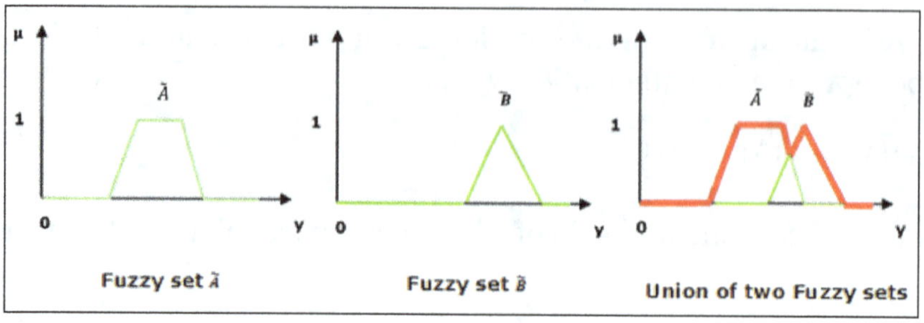

Intersection/Fuzzy 'AND'

Let us consider the following representation to understand how the Intersection/Fuzzy 'AND' relation works:

$$\mu_{\tilde{A}\cap\tilde{B}}(y) = \mu_{\tilde{A}} \wedge \mu_{\tilde{B}} \quad \forall y \in U$$

Here ∧ represents the 'min' operation.

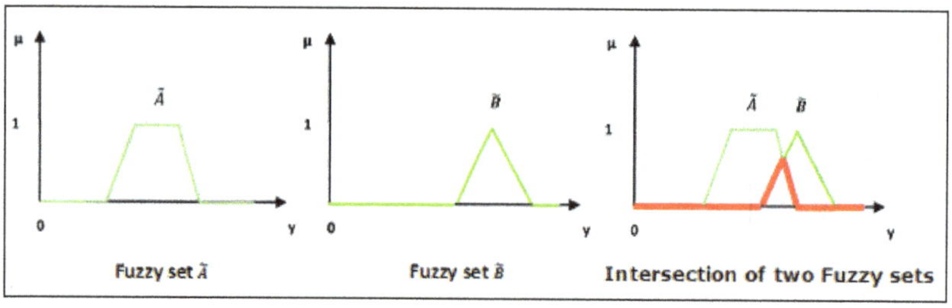

Complement/Fuzzy 'NOT'

Let us consider the following representation to understand how the Complement/ Fuzzy 'NOT' relation works:

$$\mu_{\bar{\tilde{A}}} = 1 - \mu_{\tilde{A}}(y) \quad y \in U$$

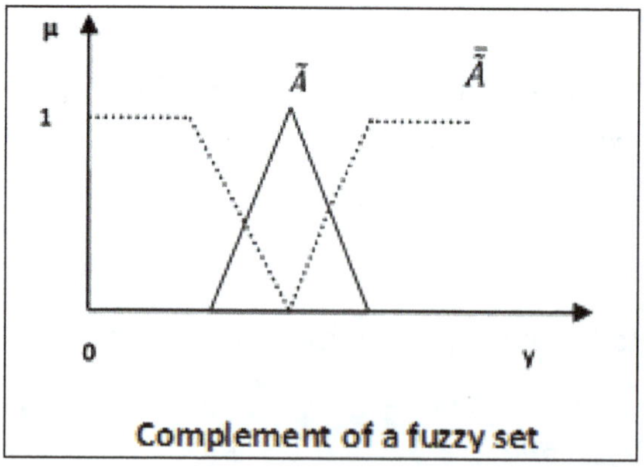

Properties of Fuzzy Sets

Commutative Property

Having two fuzzy sets $\tilde{A}$ and $\tilde{B}$, this property states:

$$\tilde{A} \cup \tilde{B} = \tilde{B} \cup \tilde{A}$$
$$\tilde{A} \cap \tilde{B} = \tilde{B} \cap \tilde{A}$$

Distributive Property

Having three fuzzy sets $\tilde{A}$, $\tilde{B}$ and $\tilde{C}$, this property states:

$$\tilde{A} \cup \left(\tilde{B} \cap \tilde{C}\right) = \left(\tilde{A} \cup \tilde{B}\right) \cap \left(\tilde{A} \cup \tilde{C}\right)$$
$$\tilde{A} \cap \left(\tilde{B} \cup \tilde{C}\right) = \left(\tilde{A} \cap \tilde{B}\right) \cup \left(\tilde{A} \cap \tilde{C}\right)$$

Idempotency Property

For any fuzzy set $\tilde{A}$, this property states:

$$\tilde{A} \cup \tilde{A} = \tilde{A}$$
$$\tilde{A} \cap \tilde{A} = \tilde{A}$$

Identity Property

For fuzzy set $\tilde{A}$ and universal set U, this property states:

$$\tilde{A} \cup \varphi = \tilde{A}$$
$$\tilde{A} \cap U = \tilde{A}$$
$$\tilde{A} \cap \varphi = \varphi$$
$$\tilde{A} \cup U = U$$

Transitive Property

Having three fuzzy sets $\tilde{A}$, $\tilde{B}$ and $\tilde{C}$, this property states:

$$\text{If } \tilde{A} \subseteq \tilde{B} \subseteq \tilde{C}, \text{then } \tilde{A} \subseteq \tilde{C}$$

Involution Property

For any fuzzy set $\tilde{A}$, this property states:

$$\bar{\bar{\tilde{A}}} \quad \tilde{A}$$

De Morgan's Law

This law plays a crucial role in proving tautologies and contradiction. This law states:

$$\overline{\tilde{A} \cap \tilde{B}} = \overline{\tilde{A} \cup \tilde{B}}$$

$$\overline{\tilde{A} \cup \tilde{B}} = \overline{\tilde{A}} \cap \overline{\tilde{B}}$$

Fuzzy Set Operations

A fuzzy set operation is an operation on fuzzy sets. These operations are generalization of crisp set operations. There is more than one possible generalization. The most widely used operations are called standard fuzzy set operations. There are three operations: fuzzy complements, fuzzy intersections, and fuzzy unions.

Standard Fuzzy Set Operations

Let A and B be fuzzy sets that A,B $\subseteq$ U, u is any element (e.g. value) in the U universe: u $\in$ U.

Standard Complement

$$\mu_{\neg A}(u) = 1 - \mu_A(u)$$

The complement is sometimes denoted by CA or AC instead of ¬A.

Standard Intersection

$$\mu_{A \cap B}(u) = \min\{\mu_A(u), \mu_B(u)\}$$

Standard Union

$$\mu_{A \cup B}(u) = \max\{\mu_A(u), \mu_B(u)\}$$

In general, the triple (i,u,n) is called De Morgan Triplet iff

- i is a t-norm,
- u is a t-conorm (aka s-norm),
- n is a strong negator.

so that for all x,y $\in$ [0, 1] the following holds true:

$$u(x,y) = n(i(n(x), n(y)))$$

(generalized De Morgan relation).

Fuzzy Complements

$\mu_A(x)$ is defined as the degree to which x belongs to A. Let $\complement A$ denote a fuzzy complement of A of type c. Then $\mu_{\complement A}(x)$ is the degree to which x belongs to $\complement A$, and the degree to which x does not belong to A. ($\mu_A(x)$ is therefore the degree to which x does not belong to $\complement A$.) Let a complement $\complement A$ be defined by a function:

$$c:[0,1]\rightarrow[0,1]$$

For all $x \in U: \mu_{\complement A}(x) = c(\mu_A(x))$

Axioms for Fuzzy Complements

Axiom c1. Boundary Condition

$$c(0) = 1 \text{ and } c(1) = 0$$

Axiom c2. Monotonicity

For all a, b $\in$ [0, 1], if a < b, then c(a) > c(b)

Axiom c3. Continuity

c is continuous function.

Axiom c4. Involutions

c is an involution, which means that c(c(a)) = a for each a $\in$ [0,1]

c is a strong negator (aka fuzzy complement).

A function c satisfying axioms c1 and c2 has at least one fixpoint a^* with $c(a^*) = a^*$, and if axiom c3 is fulfilled as well there is exactly one such fixpoint. For the standard negator c(x) = 1-x the unique fixpoint is $a^* = 0.5$.

Fuzzy Intersections

The intersection of two fuzzy sets A and B is specified in general by a binary operation on the unit interval, a function of the form:

$$i:[0,1]\times[0,1]\rightarrow[0,1].$$

For all $x \in U: \mu_{A\cap B}(x) = i[\mu_A(x), \mu_B(x)]$.

Axioms for Fuzzy Intersection

Axiom i1. Boundary Condition

$$i(a, 1) = a$$

Axiom i2. Monotonicity

$b \leq d$ implies $i(a, b) \leq i(a, d)$

Axiom i3. Commutativity

$i(a, b) = i(b, a)$

Axiom i4. Associativity

$i(a, i(b, d)) = i(i(a, b), d)$

Axiom i5. Continuity

i is a continuous function

Axiom i6. Subidempotency

$i(a, a) \leq a$

Axiom i7. Strict Monotonicity

$i(a_1, b_1) \leq i(a_2, b_2)$ if $a_1 \leq a_2$ and $b_1 \leq b_2$

Axioms i1 up to i4 define a t-norm (aka fuzzy intersection). The standard t-norm min is the only idempotent t-norm (that is, $i(a_1, a_1) = a$ for all $a \in [0,1]$).

Fuzzy Unions

The union of two fuzzy sets A and B is specified in general by a binary operation on the unit interval function of the form:

$$u : [0,1] \times [0,1] \rightarrow [0,1].$$

For all $x \in U : \mu_{A \cup B}(x) = u[\mu_A(x), \mu_B(x)]$.

Axioms for Fuzzy Union

Axiom u1. Boundary Condition

$u(a, 0) = u(0, a) = a$

Axiom u2. Monotonicity

$b \leq d$ implies $u(a, b) \leq u(a, d)$

Axiom u3. Commutativity

u(a, b) = u(b, a)

Axiom u4. Associativity

u(a, u(b, d)) = u(u(a, b), d)

Axiom u5. Continuity

u is a continuous function

Axiom u6. Superidempotency

u(a, a) ≥ a

Axiom u7. Strict Monotonicity

$$a_1 < a_2 \text{ and } b_1 < b_2 \text{ implies } u(a_1, b_1) < u(a_2, b_2)$$

Axioms u1 up to u4 define a t-conorm (aka s-norm or fuzzy intersection). The standard t-conorm max is the only idempotent t-conorm (i. e. u (a1, a1) = a for all a ∈ [0,1]).

Aggregation Operations

Aggregation operations on fuzzy sets are operations by which several fuzzy sets are combined in a desirable way to produce a single fuzzy set.

Aggregation operation on n fuzzy set (2 ≤ n) is defined by a function:

$$h : [0,1]^n \rightarrow [0,1]$$

Axioms for Aggregation Operations Fuzzy Sets

Axiom h1. Boundary Condition

h(0, 0, ..., 0) = 0 and h(1, 1, ..., 1) = one

Axiom h2. Monotonicity

For any pair $< a_1, a_2, ..., a_n >$ and $< b_1, b_2, ..., b_n >$ of n-tuples such that $a_i, b_i \in [0,1]$ for all $i \in N_n$, if $a_i \leq b_i$ for all $i \in N_n$, then $h(a_1, a_2, ..., a_n) \leq h(b_1, b_2, ..., b_n)$; that is, h is monotonic increasing in all its arguments.

Axiom h3. Continuity

h is a continuous function.

Fuzzy Classification

Fuzzy classification is the process of grouping elements into a fuzzy set whose membership function is defined by the truth value of a fuzzy propositional function.

A fuzzy class $\sim C = \{i \mid \sim \Pi(i)\}$ is defined as a fuzzy set $\sim C$ of individuals i satisfying a fuzzy classification predicate ~Π which is a fuzzy propositional function. The domain of the fuzzy class operator $\sim \{.|.\}$ is the set of variables V and the set of fuzzy propositional functions $\sim PF$, and the range is the fuzzy powerset (the set of fuzzy subsets) of this universe, $\sim P(U)$:

$$\sim \{.|.\} : V \times \sim PF \rightarrow \sim P(U)$$

A fuzzy propositional function is, analogous to, an expression containing one or more variables, such that, when values are assigned to these variables, the expression becomes a fuzzy proposition in the sense of.

Accordingly, fuzzy classification is the process of grouping individuals having the same characteristics into a *fuzzy set*. A fuzzy classification corresponds to a membership function μ that indicates whether an individual is a member of a class, given its fuzzy classification predicate $\sim \Pi$.

$$\mu : \sim PF \times U \rightarrow \sim T$$

Here, $\sim T$ is the set of fuzzy truth values (the interval between zero and one). The fuzzy classification predicate $\sim \Pi$ corresponds to a fuzzy restriction "i is R" of U, where R is a fuzzy set defined by a truth function. The degree of membership of an individual i in the fuzzy class $\sim C$ is defined by the truth value of the corresponding fuzzy predicate.

$$\mu \sim C(i) := \tau(\sim \Pi(i))$$

Classification

Intuitively, a class is a set that is defined by a certain property, and all objects having that property are elements of that class. The process of classification evaluates for a given set of objects whether they fulfill the classification property, and consequentially are a member of the corresponding class. However, this intuitive concept has some logical subtleties that need clarification.

A class logic is a logical system which supports set construction using logical predicates with the class operator $\{.|.\}$. A *class*

$$C = \{i \mid \Pi(i)\}$$

is defined as a set C of individuals i satisfying a classification predicate Π which is a propositional function. The domain of the class operator $\{.|.\}$ is the set of variables V and the set of propositional functions PF, and the range is the powerset of this universe P(U) that is, the set of possible subsets:

$$\{.|.\} : V \times PF \rightarrow P(U)$$

Here is an explanation of the logical elements that constitute this definition:

- An individual is a real object of reference.
- A universe of discourse is the set of all possible individuals considered.
- A variable $V : \rightarrow R$ is a function which maps into a predefined range R without any given function arguments: a zero-place function.
- A propositional function is "an expression containing one or more undetermined constituents, such that, when values are assigned to these constituents, the expression becomes a proposition".

In contrast, *classification* is the process of grouping individuals having the same characteristics into a set. A classification corresponds to a membership function μ that indicates whether an individual is a member of a class, given its classification predicate Π.

$$\mu : PF \times U \rightarrow T$$

The membership function maps from the set of propositional functions PF and the universe of discourse U into the set of truth values T. The membership μ of individual i in Class C is defined by the truth value τ of the classification predicate Π.

$$\mu C(i) := \tau(\Pi(i))$$

In classical logic the truth values are certain. Therefore a classification is crisp, since the truth values are either exactly true or exactly false.

FUZZY RELATIONS

A classical relation can be considered as a set of tuples, where a tuple is an ordered pair. A binary tuple is denoted by (u, v), an example of a ternary tuple is (u, v, w) and an example of n-ary tuple is $(x_1,\ldots,x_n)$.

Example: Let X be the domain of man {John, Charles, James} and Y the domain of women {Diana, Rita, Eva}, then the realtion "married to" on X × Y is, for example:

{(Charles, Diana), (John, Eva), (James, Rita)}

Definition: (classical n-ary relation) Let $X_1,\ldots,X_n$ be classical sets. The subsets of the

Cartesian product $X_1 \times \cdots \times X_n$ are called n-ary relations. If $X_1 = \cdots = X_n$ and $R \subset X^n$ then R is called an n-ary relation in X.

Let R be a binary relation in $\mathbb{R}$. Then the characteristic function of R is defined as:

$$\chi_R(u,v) = \begin{cases} 1 & \text{if } (u,v) \in R \\ 0 & \text{otherwise} \end{cases}$$

Example: Consider the following relation,

$$(u,v) \in R \Leftrightarrow u \in [a,b] \text{ and } v \in [0,c]$$

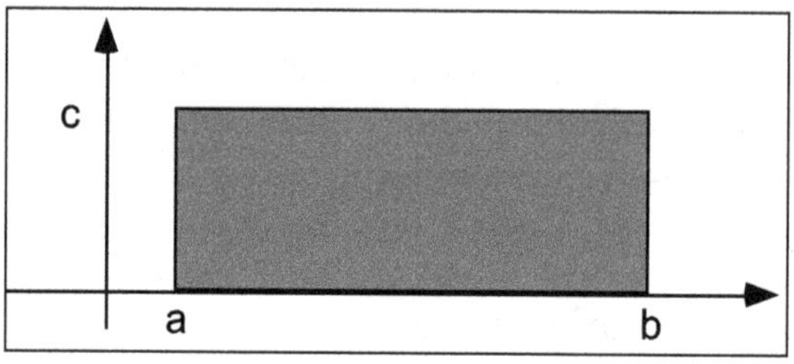

$$\chi_R(u,v) = \begin{cases} 1 & \text{if } (u,v) \in [a,b] \times [0,c] \\ 0 & \text{otherwise} \end{cases}$$

Let R be a binary relation in a classical set X. Then

Definition: (reflexivity) R is reflexive if $\forall u \in U : (u, u) \in R$

Definition: (anti-reflexivity) R is anti-reflexive if $\forall u \in U : (u, u) \notin R$

Definition: (symmetricity) R is symmetric if from $(u, v) \in R \rightarrow (v, u) \in R, \forall u, v \in U$

Definition: (anti-symmetricity) R is anti-symmetric if $(u, v) \in R$ and $(v, u) \in R$ then $u = v, \forall u, v \in U$

Definition. (transitivity) R is transitive if $(u, v) \in R$ and $(v, w) \in R$ then $(u, w) \in R, \forall u, v, w \in U$

Example: Consider the classical inequality relations on the real line $\mathbb{R}$. It is clear that $\leq$ is reflexive, anti-symmetric and transitive, $<$ is anti-reflexive, anti-symmetric and transitive.

Other important properties of binary relations are:

Property: (equivalence) R is an equivalence relation if, R is reflexive, symmetric and transitive.

Property: (partial order) R is a partial order relation if it is reflexive, anti-symmetric and transitive.

Property: (total order) R is a total order relation if it is partial order and ∀u, v ∈ R, (u, v) ∈ R or (v, u) ∈ R hold.

Example: Let us consider the binary relation "subset of". It is clear that we have a partial order relation.

The relation ≤ on natural numbers is a total order relation.

Consider the relation "mod 3" on natural numbers

$$\{(m, n) \mid (n - m) \bmod 3 \equiv 0\}$$

This is an equivalence relation.

Definition: Let X and Y be nonempty sets. A fuzzy relation R is a fuzzy subset of X × Y.

In other words, $R \in \mathcal{F}(X \times Y)$

If X = Y then we say that R is a binary fuzzy relation in X.

Let R be a binary fuzzy relation on $\mathbb{R}$. Then R(u, v) is interpreted as the degree of membership of the ordered pair (u, v) in R.

Example: A simple example of a binary fuzzy relation on:

$$U = \{1, 2, 3\},$$

called "approximately equal" can be defined as:

$$R(1, 1) = R(2, 2) = R(3, 3) = 1$$

$$R(1, 2) = R(2, 1) = R(2, 3) = R(3, 2) = 0.8$$

$$R(1, 3) = R(3, 1) = 0.3$$

The membership function of R is given by:

$$R(u,v) = \begin{cases} 1 & if\ u = v \\ 0.8 & if\ |u-v| = 1 \\ 0.3 & if\ |u-v| = 2 \end{cases}$$

In matrix notation it can be represented as:

$$\begin{pmatrix} & 1 & 2 & 3 \\ 1 & 1 & 0.8 & 0.3 \\ 2 & 0.8 & 1 & 0.8 \\ 3 & 0.3 & 0.8 & 1 \end{pmatrix}$$

Operations on Fuzzy Relations

Fuzzy relations are very important because they can describe interactions between variables. Let R and S be two binary fuzzy relations on X × Y.

Definition: The intersection of R and S is defined by:

$$(R \wedge S)(u,\ v) = \ \min\{R(u,\ v),\ S(u,v)\}.$$

Note that $R:\ X \times Y \rightarrow [0,\ 1]$, i.e. R the domain of R is the whole Cartesian product X × Y.

Definition: The union of R and S is defined by:

$$(R \vee S)(u,v) = \max\{R(u,v),\ S(u,v)\}$$

Example: Let us define two binary relations:

R = "x is considerable larger than y"

$$= \begin{pmatrix} & y_1 & y_2 & y_3 & y_4 \\ x_1 & 0.8 & 0.1 & 0.1 & 0.7 \\ x_2 & 0 & 0.8 & 0 & 0 \\ x_3 & 0.9 & 1 & 0.7 & 0.8 \end{pmatrix}$$

S = "x is very close to y"

$$= \begin{pmatrix} & y_1 & y_2 & y_3 & y_4 \\ x_1 & 0.4 & 0 & 0.9 & 0.6 \\ x_2 & 0.9 & 0.4 & 0.5 & 0.7 \\ x_3 & 0.3 & 0 & 0.8 & 0.5 \end{pmatrix}$$

The intersection of R and S means that "x is considerable larger than y" and "x is very close to y".

$$R(\wedge S)(x,y) = \begin{pmatrix} & y_1 & y_2 & y_3 & y_4 \\ x_1 & 0.4 & 0 & 0.1 & 0.6 \\ x_2 & 0 & 0.4 & 0 & 0 \\ x_3 & 0.3 & 0 & 0.7 & 0.5 \end{pmatrix}$$

The union of R and S means that x is considerable larger than "y or x is very close to y".

$$R(\wedge S)(x,y)=\begin{pmatrix} & y_1 & y_2 & y_3 & y_4 \\ x_1 & 0.8 & 0 & 0.9 & 0.7 \\ x_2 & 0.9 & 0.8 & 0.5 & 0.7 \\ x_3 & 0.9 & 1 & 0.8 & 0.8 \end{pmatrix}$$

Consider a classical relation R on $\mathbb{R}$.

$$R(u,v)=\begin{cases} 1 & if\ (u,v)\in[a,b]\times[0,c] \\ 0 & otherwise \end{cases}$$

It is clear that the projection (or shadow) of R on the X-axis is the closed interval [a, b] and its projection on the Y -axis is [0, c].

If R is a classical relation in X × Y then:

$$\Pi_X=\{x\in X\mid \exists y\in Y:(x,y)\in R\}$$

$$\Pi_Y=\{y\in Y\mid \exists x\in X:(x,y)\in R\}$$

where Π_X denotes projection on X and Π_Y denotes projection on Y.

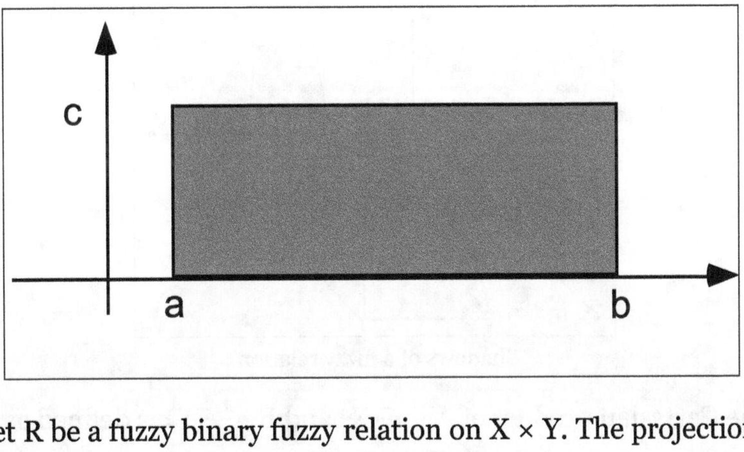

Definition: Let R be a fuzzy binary fuzzy relation on X × Y. The projection of R on X is defined as

$$\Pi_X(x)=\sup\{R(x,\ y)\mid y\in Y\}$$

and the projection of R on Y is defined as:

$$\Pi_Y(y)=\sup\{R(x,\ y)\mid x\in X\}$$

Example: Consider the relation,

R = “x is considerable larger than y”

$$= \begin{pmatrix} & y_1 & y_2 & y_3 & y_4 \\ x_1 & 0.8 & 0.1 & 0.1 & 0.7 \\ x_2 & 0 & 0.8 & 0 & 0 \\ x_3 & 0.9 & 1 & 0.7 & 0.8 \end{pmatrix}$$

then the projection on X means that:

- x_1 is assigned the highest membership degree from the tuples $(x_1, y_1),(x_1, y_2),(x_1, y_3),(x_1, y_4)$, i.e. $\Pi_X(x_1)=1$, which is the maximum of the first row.
- x_2 is assigned the highest membership degree from the tuples $(x_2, y_1),(x_2, y_2),(x_2, y_3),(x_2, y_4)$, i.e. $\Pi_X(x_2)=0.8$, which is the maximum of the second row.
- x_3 is assigned the highest membership degree from the tuples $(x_3, y_1),(x_3, y_2),(x_3, y_3),(x_3, y_4)$, i.e. $\Pi_X(x_3)=1$, which is the maximum of the third row.

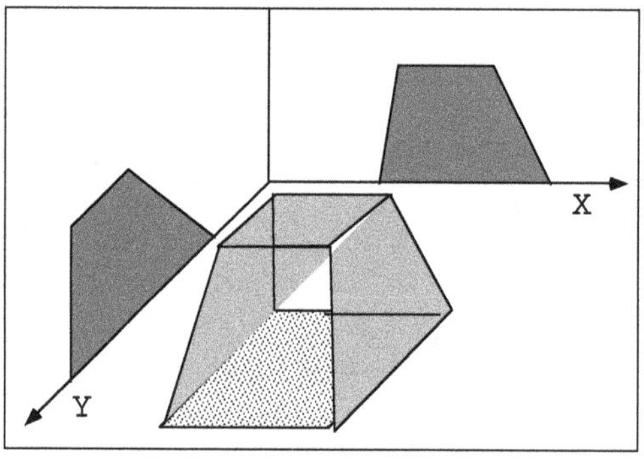

Shadows of a fuzzy relation.

Definition: The Cartesian product of A ∈ $\mathcal{F}$(X) and B ∈ $\mathcal{F}$(Y) is defined as:

$$(A\times B)(u,v)=\min\{A(u),\ B(v)\}.$$

for all u ∈ X and v ∈ Y.

It is clear that the Cartesian product of two fuzzy sets is a fuzzy relation in X × Y.

If A and B are normal then $\Pi_Y(A\times B)=B$ and $\Pi_X(A\times B)=A$.

Really,

$$\Pi_X(x) = \sup\{(A \times B)(x, y) \mid y\}$$
$$= \sup\{A(x) \wedge B(y) \mid y\} =$$
$$\min\{A(x), \sup\{B(y)\} \mid y\}$$
$$= \min\{A(x), 1\} = A(x).$$

Definition: The sup-min composition of a fuzzy set C ∈ $\mathcal{F}$(X) and a fuzzy relation R ∈ (X × Y) is defined as:

$$(C \circ R)(y) = \sup_{x \in X} \min\{C(x), R(x, y)\}$$

for all y ∈ Y.

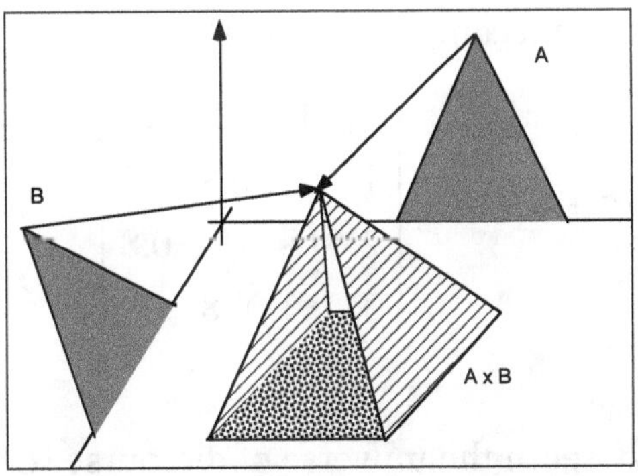

The composition of a fuzzy set C and a fuzzy relation R can be considered as the shadow of the relation R on the fuzzy set C.

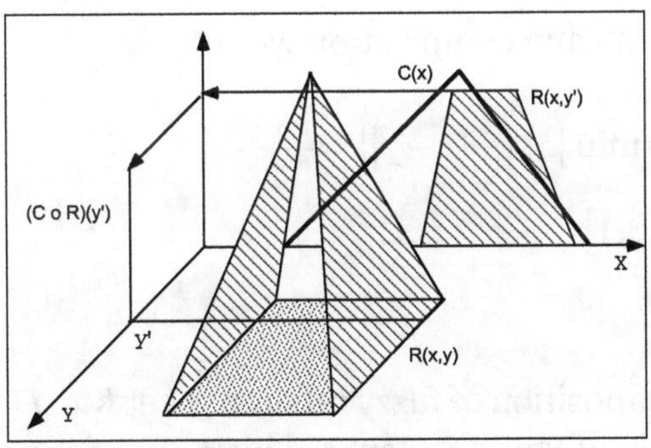

Example: Let A and B be fuzzy numbers and let,

R = A × B

a fuzzy relation.

Observe the following property of composition

$$A \cdot R = A \cdot (A \times B) = A,$$

$$B \cdot R = B \cdot (A \times B) = B$$

Example: Let C be a fuzzy set in the universe of discourse {1, 2, 3} and let R be a binary fuzzy relation in {1, 2, 3}. Assume that:

$$C = 0.2/1 + 1/2 + 0.2/3$$

And

$$R = \begin{pmatrix} & 1 & 2 & 3 \\ 1 & 1 & 0.8 & 0.3 \\ 2 & 0.8 & 1 & 0.8 \\ 3 & 0.3 & 0.8 & 1 \end{pmatrix}$$

Using the definition of sup-min composition we get:

$$C \cdot R = (0.2/1 + 1/2 + 0.2/3) \cdot \begin{pmatrix} & 1 & 2 & 3 \\ 1 & 1 & 0.8 & 0.3 \\ 2 & 0.8 & 1 & 0.8 \\ 3 & 0.3 & 0.8 & 1 \end{pmatrix} =$$

$$0.8/1 + 1/2 + 0.8/3.$$

Example: Let C be a fuzzy set in the universe of discourse [0, 1] and let R be a binary fuzzy relation in [0, 1]. Assume that C(x) = x and,

$$R(x, y) = 1 - |x - y|.$$

Using the definition of sup-min composition we get

$$(C \cdot R)(y) = \sup_{x \in [0,1]} \min\{x, 1 - |x - y|\} = \frac{1 + y}{2}$$

for all y ∈ [0, 1].

Definition: (sup-min composition of fuzzy relations) Let $R \in \mathcal{F}(X \times Y)$ and $S \in \mathcal{F}(Y \times Z)$. The sup-min composition of R and S, denoted by R · S is defined as

$$(R \cdot S)(u, w) = \sup_{v \in Y} \min\{R(u, v),\ S(v,\ w)\}$$

It is clear that R · S is a binary fuzzy relation in X × Z.

Example: Consider two fuzzy relations,

R = "x is considerable larger than y"

$$= \begin{pmatrix} & y_1 & y_2 & y_3 & y_4 \\ x_1 & 0.8 & 0.1 & 0.1 & 0.7 \\ x_2 & 0 & 0.8 & 0 & 0 \\ x_3 & 0.9 & 1 & 0.7 & 0.8 \end{pmatrix}$$

$$S = \text{"y is very close to z"} = \begin{pmatrix} & z_1 & z_2 & z_3 \\ y_1 & 0.4 & 0.9 & 0.3 \\ y_2 & 0 & 0.4 & 0 \\ y_3 & 0.9 & 0.5 & 0.8 \\ y_4 & 0.6 & 0.7 & 0.5 \end{pmatrix}$$

Then their composition is:

$$R \cdot S - \begin{pmatrix} & z_1 & z_2 & z_3 \\ x_1 & 0.6 & 0.8 & 0.5 \\ x_2 & 0 & 0.4 & 0 \\ x_3 & 0.7 & 0.9 & 0.7 \end{pmatrix}$$

Formally,

$$\begin{pmatrix} & y_1 & y_2 & y_3 & y_4 \\ x_1 & 0.8 & 0.1 & 0.1 & 0.7 \\ x_2 & 0 & 0.8 & 0 & 0 \\ x_3 & 0.9 & 1 & 0.7 & 0.7 \end{pmatrix} \cdot \begin{pmatrix} & z_1 & z_2 & z_3 \\ y_1 & 0.4 & 0.9 & 0.3 \\ y_2 & 0 & 0.4 & 0 \\ y_3 & 0.9 & 0.5 & 0.8 \\ y_4 & 0.6 & 0.7 & 0.5 \end{pmatrix} =$$

$$\begin{pmatrix} & z_1 & z_2 & z_3 \\ x_1 & 0.6 & 0.8 & 0.5 \\ x_2 & 0 & 0.4 & 0 \\ x_3 & 0.7 & 0.9 & 0.7 \end{pmatrix}$$

i.e., the composition of R and S is nothing else, but the classical product of the matrices R and S with the difference that instead of addition we use maximum and instead of multiplication we use minimum operator.

FUZZY RULE

Fuzzy rules are used within fuzzy logic systems to infer an output based on input variables. Modus ponens and modus tollens are the most important rules of inference. A modus ponens rule is in the form:

Premise: x is A.

Implication: IF x is A THEN y is B.

Consequent: y is B.

In crisp logic, the premise x is A can only be true or false. However, in a fuzzy rule, the premise x is A and the consequent y is B can be true to a degree, instead of entirely true or entirely false. This is achieved by representing the linguistic variables A and B using fuzzy sets. In a fuzzy rule, modus ponens is extended to generalised modus ponens:

Premise: x is A*.

Implication: IF x is A THEN y is B.

Consequent: y is B*.

The key difference is that the premise x is A can be only partially true. As a result, the consequent y is B is also partially true. Truth is represented as a real number between 0 and 1, where 0 is false and 1 is true.

Comparison between Boolean and Fuzzy Logic Rules

As an example, consider a rule used to control a three-speed fan. A binary IF-THEN statement may be:

IF temperature $\geq$30.

THEN fan speed is 3.

The disadvantage of this rule is that it uses a strict temperature as a threshold, but the user may want the fan to still function at this speed when temperature = 29.9. A fuzzy IF-THEN statement may be:

IF temperature is hot.

THEN fan speed is fast.

where hot and fast are described using fuzzy sets.

Fuzzy Rule Connectors

Rules can connect multiple variables through fuzzy set operations using t-norms and t-conorms.

T-norms are used as an AND connector. For example.

IF temperature is hot AND humidity is high.

THEN fan speed is fast.

A degree of truth is assigned to temperature is hot and to humidity is high. The result of a t-norm operation on these two degrees is used as the degree of truth that fan speed is fast.

T-conorms are used as an OR connector. For example.

IF temperature is hot OR humidity is high.

THEN fan speed is fast.

The result of a t-conorm operation on these two degrees is used as the degree of truth that fan speed is fast.

The complement of a fuzzy set is used as a negator. For example,

IF temperature is NOT hot.

THEN fan speed is slow.

The fuzzy set not hot is the complement of hot. The degree of truth assigned to temperature is not hot is used as the degree of truth that fan speed is slow.

T-conorms are less commonly used as rules can be represented by AND and OR connectors exclusively.

FUZZY ASSOCIATIVE MATRIX

A fuzzy associative matrix expresses fuzzy logic rules in tabular form. These rules usually take two variables as input, mapping cleanly to a two-dimensional matrix, although theoretically a matrix of any number of dimensions is possible.

Suppose a professional is tasked with writing fuzzy logic rules for a video game monster. In the game being built, entities have two variables: hit points (HP) and firepower (FP):

HP/FP	Very low HP	Low HP	Medium HP	High HP	Very high HP
Very weak FP	Retreat!	Retreat!	Defend	Defend	Defend
Weak FP	Retreat!	Defend	Defend	Attack	Attack
Medium FP	Retreat!	Defend	Attack	Attack	Full attack!
High FP	Retreat!	Defend	Attack	Attack	Full attack!
Very high FP	Defend	Attack	Attack	Full attack!	Full attack!

This translates to:

```
IF MonsterHP IS VeryLowHP AND MonsterFP IS VeryWeakFP THEN Retreat

IF MonsterHP IS LowHP AND MonsterFP IS VeryWeakFP THEN Retreat

IF MonsterHP IS MediumHP AND MonsterFP is VeryWeakFP THEN Defend
```

Multiple rules can fire at once, and often will, because the distinction between "very low" and "low" is fuzzy. If it is more "very low" than it is low, then the "very low" rule will generate a stronger response. The program will evaluate all the rules that fire and use an appropriate defuzzification method to generate its actual response.

An implementation of this system might use either the matrix or the explicit IF/THEN form. The matrix makes it easy to visualize the system, but it also makes it impossible to add a third variable just for one rule, so it is less flexible.

There is no inherent pattern in the matrix. It appears as if the rules were just made up, and indeed they were. This is both a strength and a weakness of fuzzy logic in general. It is often impractical or impossible to find an exact set of rules or formulae for dealing with a specific situation. For a sufficiently complex game, a mathematician would not be able to study the system and figure out a mathematically accurate set of rules. However, this weakness is intrinsic to the realities of the situation, not of fuzzy logic itself. The strength of the system is that even if one of the rules is wrong, even greatly wrong, other rules that are correct are likely to fire as well and they may compensate for the error.

This does not mean a fuzzy system should be sloppy. Depending on the system, it might get away with being sloppy, but it will underperform. While the rules are fairly arbitrary, they should be chosen carefully. If possible, an expert should decide on the rules, and the sets and rules should be tested vigorously and refined as needed. In this way, a fuzzy system is like an expert system.

FUZZY MEASURE THEORY

Fuzzy measure theory considers generalized measures in which the additive property is replaced by the weaker property of monotonicity. The central concept of fuzzy measure theory is the fuzzy measure (also *capacity*) which was introduced by Choquet in 1953 and independently defined by Sugeno in 1974 in the context of fuzzy integrals. There exists a number of different classes of fuzzy measures including plausibility/belief measures; possibility/necessity measures; and probability measures which are a subset of classical measures.

Let $\mathbf{X}$ be a universe of discourse, C be a class of subsets of $\mathbf{X}$, and $E, F \in C$. A function $g : C \rightarrow \mathbb{R}$ where

$$\varnothing \in \mathcal{C} \Rightarrow g(\varnothing) = 0.$$

$$E \subseteq F \Rightarrow g(E) \leq g(F).$$

is called a *fuzzy measure*. A fuzzy measure is called *normalized* or *regular* if $g(\mathbf{X}) = 1$.

Properties of Fuzzy Measures

A fuzzy measure is:

- Additive if for any $E, F \in \mathcal{C}$ such that $E \cap F = \varnothing$, we have

  $g(E \cup F) = g(E) + g(F)$;
- Supermodular if for any $E, F \in \mathcal{C}$, we have

  $g(E \cup F) + g(E \cap F) \geq g(E) + g(F)$;
- Submodular if for any $E, F \in \mathcal{C}$, we have

  $g(E \cup F) + g(E \cap F) \leq g(E) + g(F)$;
- Superadditive if for any $E, F \in \mathcal{C}$ such that $E \cap F = \varnothing$, we have

  $g(E \cup F) \geq g(E) + g(F)$;
- Subadditive if for any $E, F \in \mathcal{C}$ such that $E \cap F = \varnothing$, we have

  $g(E \cup F) \leq g(E) + g(F)$;
- Symmetric if for any $E, F \in \mathcal{C}$, we have

  $|E| = |F|$ implies $g(E) = g(F)$;
- Boolean if for any $E \in \mathcal{C}$, we have

  $g(E) = 0$ or $g(E) = 1$.

Understanding the properties of fuzzy measures is useful in application. When a fuzzy measure is used to define a function such as the Sugeno integral or Choquet integral, these properties will be crucial in understanding the function's behavior. For instance, the Choquet integral with respect to an additive fuzzy measure reduces to the Lebesgue integral. In discrete cases, a symmetric fuzzy measure will result in the ordered weighted averaging (OWA) operator. Submodular fuzzy measures result in convex functions, while supermodular fuzzy measures result in concave functions when used to define a Choquet integral.

Möbius Representation

Let g be a fuzzy measure, the Möbius representation of g is given by the set function M, where for every $E, F \subseteq X$,

$$M(E) = \sum_{F \subseteq E} (-1)^{|E \setminus F|} g(F).$$

The equivalent axioms in Möbius representation are:

$$M(\varnothing) = 0$$

$$\sum_{F \subseteq E | i \in F} M(F) \geq 0 \text{ for all } E \subseteq \mathbf{X} \text{ and all } i \in E$$

A fuzzy measure in Möbius representation M is called normalized if $\sum_{E \subseteq \mathbf{X}} M(E) = 1$.

Möbius representation can be used to give an indication of which subsets of **X** interact with one another. For instance, an additive fuzzy measure has Möbius values all equal to zero except for singletons. The fuzzy measure g in standard representation can be recovered from the Möbius form using the Zeta transform:

$$g(E) = \sum_{F \subseteq E} M(F), \forall E \subseteq \mathbf{X}.$$

Simplification Assumptions for Fuzzy Measures

Fuzzy measures are defined on a semiring of sets or monotone class which may be as granular as the power set of X, and even in discrete cases the number of variables can as large as $2^{|X|}$. For this reason, in the context of multi-criteria decision analysis and other disciplines, simplification assumptions on the fuzzy measure have been introduced so that it is less computationally expensive to determine and use. For instance, when it is assumed the fuzzy measure is *additive*, it will hold that $g(E) = \sum_{i \in E} g(\{i\})$ and the values of the fuzzy measure can be evaluated from the values on X. Similarly, a *symmetric* fuzzy measure is defined uniquely by |X| values. Two important fuzzy measures that can be used are the Sugeno- or λ -fuzzy measure and k-additive measures, introduced by Sugeno and Grabisch respectively.

Sugeno λ-measure

The Sugeno λ -measure is a special case of fuzzy measures defined iteratively. It has the following definition:

Let $X = \{x_1, ..., x_n\}$ be a finite set and let $\lambda \in (-1, +\infty)$ A Sugeno λ -measure is a function $g : 2^X \rightarrow [0,1]$ such that,

$g(X)=1$.

If $A,B \subseteq \mathbf{X}$ (alternatively $A,B \in 2^{\mathbf{X}}$) with $A \cap B = \varnothing$ then
$g(A \cup B) = g(A) + g(B) + \lambda g(A)g(B)$.

As a convention, the value of g at a singleton set $\{x_i\}$ is called a density and is denoted by $g_i = g(\{x_i\})$. In addition, we have that λ satisfies the property:

$$\lambda + 1 = \prod_{i=1}^{n} (1 + \lambda g_i).$$

Tahani and Keller as well as Wang and Klir have showed that once the densities are known, it is possible to use the previous polynomial to obtain the values of λ uniquely.

k-additive Fuzzy Measure

The *k*-additive fuzzy measure limits the interaction between the subsets $E \subseteq X$ to size $|E| = k$. This drastically reduces the number of variables needed to define the fuzzy measure, and as *k* can be anything from 1 (in which case the fuzzy measure is additive) to **X**, it allows for a compromise between modelling ability and simplicity.

A discrete fuzzy measure *g* on a set X is called *k-additive* $(1 \le k \le |\mathbf{X}|)$ if its Möbius representation verifies $M(E) = 0$, whenever $|E| > k$ for any $E \subseteq \mathbf{X}$, and there exists a subset *F* with *k* elements such that $M(F) \neq 0$.

Shapley and Interaction Indices

In game theory, the Shapley value or Shapley index is used to indicate the weight of a game. Shapley values can be calculated for fuzzy measures in order to give some indication of the importance of each singleton. In the case of additive fuzzy measures, the Shapley value will be the same as each singleton.

For a given fuzzy measure *g*, and $|\mathbf{X}| = n$, the Shapley index for every $i, \ldots, n \in X$ is:

$$\phi(i) = \sum_{E \subseteq \mathbf{X} \setminus \{i\}} \frac{(n - |E| - 1)!\,|E|!}{n!} [g(E \cup \{i\}) - g(E)].$$

The Shapley value is the vector $\phi(g) = (\psi(1), \ldots, \psi(n))$.

FUZZY FINITE ELEMENT METHOD

The fuzzy finite element method combines the well-established finite element method with the concept of fuzzy numbers, the latter being a special case of a fuzzy set. The advantage of using fuzzy numbers instead of real numbers lies in the incorporation of uncertainty (on material properties, parameters, geometry, initial conditions, etc.) in the finite element analysis.

One way to establish a fuzzy finite element (FE) analysis is to use existing FE software (in-house or commercial) as an inner-level module to compute a deterministic result, and to add an outer-level loop to handle the fuzziness (uncertainty). This outer-level loop comes down to solving an optimization problem. If the inner-level deterministic module produces monotonic behavior with respect to the input variables, then the outer-level optimization problem is greatly simplified, since in this case the extrema will be located at the vertices of the domain.

Traditional finite element method is a well-established method to solve various problems of science and engineering. Different authors have used various methods to solve governing differential equation of heat conduction problem. In this study, heat conduction in a circular rod has been considered which is made up of two different materials viz. aluminum and copper. In earlier studies parameters in the differential equation have been taken as fixed numbers which actually may not. Those parameters are found in general by some measurements or experiments. So the material properties are actually uncertain and may be considered to vary in an interval or as fuzzy and in that case complex interval arithmetic or fuzzy arithmetic has to be considered in the analysis. Hence interval/fuzzy arithmetic is applied in the finite element method to solve a steady state heat conduction problem. Application of fuzzy finite element method in the said problem gives fuzzy system of linear equations in general.

STANDARD ADDITIVE MODEL FUZZY INFERENCE

The Standard Additive Model (SAM) is a generalized inference model proposed and analysed by Kosko. In SAM, an additive fuzzy system, F stores m fuzzy rules of the form "if x = A_j then y = B_j" and computes the output F(x) as the centeroid (or mode) of the summed and partially fired then-part fuzzy sets $B_j^{'}$. Each input x fires all the rules to some degree and in parallel. So the additive fuzzy system $F:\Re^{n} \rightarrow \Re^{p}$ acts as an associative processor or global fuzzy associative memory (FAM). The feed-forward standard additive model (SAM) is shown in figure. In the SAM case, the fired then-part set

$B_j^{'}$ is just the fit product aj (x) Bj. The additive structure comes from the sum B offired then-part sets.

$$B = \sum_{j=1}^{m} w_j B_j^{'} = \sum_{j=1}^{m} w_j a_j(x) B_j$$

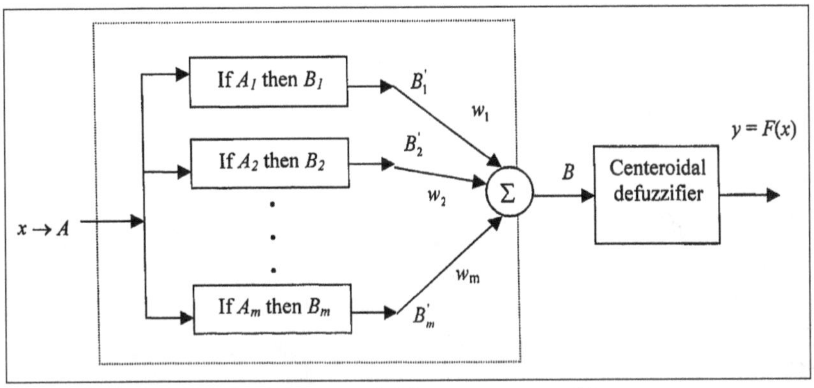

Standard Additive Model fuzzy system architecture.

Real input x, fires the if-part of all m rules, in parallel and to some degree (most-to-zero degree). Then the system weights the then-parts to give the new fuzzy sets B~ and then sums these to form the output set B.

The system "defuzzifies" B to give the output, y = F (x) by taking the centeroid of B. The sum over the m rules that map fuzzy subsets, $A_j \subset \Re"$ to fuzzy subsets: $B_j \subset \Re^P$. The fit (fuzzy unit) value $a_j(x)$ states the degree to which the input x belongs to the if part set A_j So $a_j : \Re^n \rightarrow [0,1]$ and $b_j : \Re^p \rightarrow [0,1]$ define set functions. The rule weights w_j can be any real non-negative numbers. The sum in (3.1) leads to a SAM system. The rule weights w_j can scale each term in the sum and gives an extra term for a learning system to tune. Further, we can map B to F(x) by centeroid which gives the fuzzy system F the structure of a conditional expectation and acts as an optimal nonlinear approximator in the mean squared sense. Also, B can be mapped to F(x) by the mode of B which makes the fuzzy system F a MAP or maximum a posteriori estimate. The centroid choice F(x) = centroid(B) uses all the information in the output set B and hence is optimal in the Bayesian sense. We can now state and prove the simplest form of SAM Theorem. It shows that a fuzzy system maps input x to F(x) as a convex sum of the m then-part set centeroids c_j. These centeroids can change with each x as in a generalized SAM or can change with time as in an adaptive SAM. The SAM Theorem decomposes the global centeroid of B in to convex sum of local centeroids. The m coefficients $p_1(x), p_2(x).....p_m(x)$ are convex in that each term is non-negative i.e., $p_j(x) \geq 0$ and, they sum to unity: $\sum_{j=1}^{m} P_j(x) = 1.$

So the weights define a discrete probability density function for each, x.

SAM Theorem

Theorem:

Suppose the fuzzy system $F:\Re^n \rightarrow \Re^p$ is a standard additive model: F(x) = Centeroid(B) = Centeroid $\left(\sum_{j=1}^{m} w_j a_j(x) B_j\right)$, then F(x) is a convex sum of the m then-part set centeroids:

$$F(x)=\frac{\sum_{j=1}^{m} w_j a_j(x) V_j c_j}{\sum_{j=1}^{m} w_j a_j(x) V_j}$$

$$=\sum_{j=1}^{m} p_j(x) c_j$$

The convex coefficients or discrete probability weights $p_1(x), p_2(x), \ldots p_m(x)$ depend on the input x through the following ratios are given by:

$$p_j(x)=\frac{w_j a_j(x) V_j}{\sum_{k=1}^{m} w_k a_k(x) V_k}$$; V_j is the finite positive volume and c) is the centeroid of then-part set B_j:

$$V_j=\int_{\Re^p} b_j(y_1, \ldots y_p) dy_1 \ldots dy_p > 0$$

$$c_j=\frac{\int_{\Re^p} y\, b_j(y_1 \ldots y_p) dy_1 \ldots dy_p}{\int_{\Re^p} b_j(y_1 \ldots y_p) dy_1 \ldots dy_p}$$

For the popular scalar case, the above equations reduce to:

$$V_j=\int_{-\infty}^{\infty} b_j(y) dy$$

$$c_j=\frac{\int_{-\infty}^{\infty} y\, b_j(y) dy}{\int_{-\infty}^{\infty} b_j(y) dy}$$

Proof:

The Theorem follows by expanding the centeroid of B(x) and invoking the SAM assumption $B = \sum_{j=1}^{m} w_j B_j^{'} = \sum_{j=1}^{m} w_j a_j(x) B_j$) to rearrange terms:

$$F(x) = Centroid(B) = \frac{\int_{\Re^p} y b(y) dy}{\int_{\Re^p} b(y) dy} = \frac{\int_{\Re^p} y \sum_{j=1}^{m} w_j b_j^{'}(y) dy}{\int_{\Re^p} \sum_{j=1}^{m} w_j b_j^{'}(y) dy} = \frac{\int_{\Re^p} y \sum_{j=1}^{m} w_j a_j(x) b_j(y) dy}{\int_{\Re^p} \sum_{j=1}^{m} w_j a_j(x) b_j(y) dy}$$

$$= \frac{\sum_{j=1}^{m} w_j a_j(x) \int_{\Re^p} y b_j(y) dy}{\sum_{j=1}^{m} w_j a_j(x) \int_{\Re^p} b_j(y) dy} = \frac{\sum_{j=1}^{m} w_j a_j(x) V_j \frac{\int_{\Re^p} y b_j(y) dy}{V_j}}{\sum_{j=1}^{m} w_j a_j(x) V_j} = \frac{\sum_{j=1}^{m} w_j a_j(x) V_j c_j}{\sum_{j=1}^{m} w_j a_j(x) V_j}$$

We can compute the volumes V_j and centeroids c_j in $F(x) = \frac{\sum_{j=1}^{m} w_j a_j(x) V_j c_j}{\sum_{j=1}^{m} w_j a_j(x) V_j}$ in advance. They change only when the system learns or tunes its rules or when the user varies them to hand tune the system. Then the then-part sets B_j can take the form triangular or trapezoids or bell curves, linear functions of premise variables as in Takagi-Sugeno inference or more generally even nonlinear functions of premise variables.

Properties of SAM

- The SAM structure in $F(x) = \frac{\sum_{j=1}^{m} w_j a_j(x) V_j c_j}{\sum_{j=1}^{m} w_j a_j(x) V_j}$, if we replace the then-part fuzzy sets B_j with rectangular or non-fuzzy sets R_j that have the same volume V_j and centeroid C_j, this does not change the output F(x) but it can change the variance or uncertainty of the output. A symmetric set B_j has the same centeroid no matter how wide or thin the set may be. But a thicker set codes less certain knowledge and thus so does a thicker rule patch. Most applications ignore this second level of output uncertainty.

- In $F(x) = \frac{\sum_{j=1}^{m} w_j a_j(x) V_j c_j}{\sum_{j=1}^{m} w_j a_j(x) V_j}$ we can normalize the volumes to unity so that the terms V_j do not appear in $F(x)=\frac{\sum_{j=1}^{m} w_j a_j(x) V_j c_j}{\sum_{j=1}^{m} w_j a_j(x) V_j}$. Otherwise the volumes act as rule weights. The larger V_j the more it forces the global output F(x) to act like the local rule output C_j. This is shown below:

$$V_j \lim_{V_j \longrightarrow \infty} F(x) = \lim_{V_j \longrightarrow \infty} \frac{\sum_{k=1}^{m} a_k V_k c_k}{\sum_{k=1}^{m} a_k V_k}$$

$$= \lim_{V_j \longrightarrow \infty} \frac{a_j V_j c_j + C}{a_j V_j + d}$$

$$= c_j$$

- Learning schemes may also require that F be differentiable. This holds just in case all the if-part sets aix) are differentiable. So the fuzzy system is only piece-wise differentiable if the if-part sets are triangles, trapezoids, rectangles or other curves with "corners".

- $\sum_{j=1}^{m} p_j(x) = 1$:implies that $c_t \le F(x) \le c_m$

This holds for all densities p(x) for all centeroidal additive systems.

SAM and the Mamdani Inferences

Combining If-part Set Functions

The SAM Theorem allows us to pick arbitrary joint if-part set functions $a_j : \Re^n \to [0,1]$. In practice, we often work with n scalar set-function factors $a_j^1, a_j^2, ...a_j^n$. These factors $a_j^i : \Re \to [0,1]$ are often the familiar triangles, trapezoids, bell curves etc. The popular ways to combine the if-part sets is with pairwise minimum or product in place of "min".

$$a_j(x) = a_j^1(x_1) \wedge a_j^2(x_2) \wedge ..... \wedge a_j^n(x_n)$$

$$a_j(x) = \prod_{i=1}^{n} a_j^i(x_i)$$

where, minimum and product represents logical AND. We may use OR also in if-part sets, but this occurs only seldom in practice. Mamdani inference uses $a_j(x) = a_j^1(x_1) \wedge a_j^2(x_2) \wedge \ldots\ldots \wedge a_j^n(x_n)$ for combining if-part set functions.

SAMtheoremsupportsbothcombinersaboveLe., $a_j(x) = a_j^1(x_1) \wedge a_j^2(x_2) \wedge \ldots\ldots \wedge a_j^n(x_n)$ & $a_j(x) = \prod_{i=1}^{n} a_j^i(x_i)$. But, the product if-part combiner has many advantages over the min combiner. The most important advantage for the product if-part combiner is as follows. The product combiner does not ignore information, as does the min combiner. Information loss due to min combiner is known as min clipping.

Combining Fired Then-part Sets

We repeat that SAM combines the fired then part sets by convex weighted addition. i.e.,

$$B = \sum_{j=1}^{m} a_j(x) B_j \qquad \text{(SAM)}$$

Where as, other popular inferences like Mamdani inference carryout pair-wise maxima offixed then-part sets.

$$\text{i.e., } B = \bigvee_{j=1}^{m} a_j(x) B_j = \bigvee_{j=1}^{m} B_j \qquad \text{(Mamdani)}$$

The result B of $B = \bigvee_{j=1}^{m} a_j(x) B_j = \bigvee_{j=1}^{m} B_j$ (Mamdani) which uses max combiner for then-part sets make the B as a envelop or silhouette ofthe "fired" then-part sets B_j'. There is no easy way to compute the centroid of B in the formula F(x) = centeroid (B). The nature of B in this case is shown in figure.

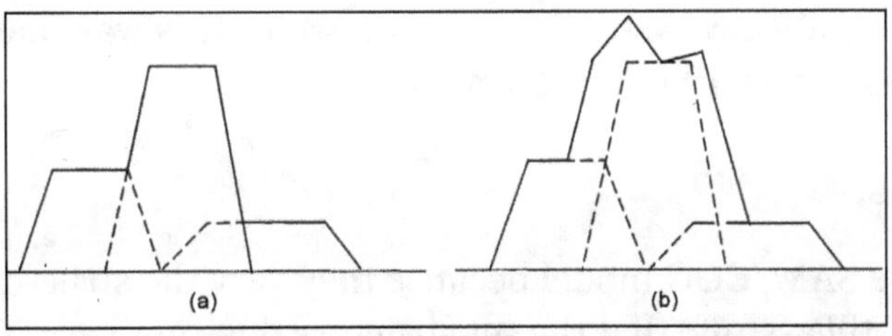

Max versus Sum combination offired then-part sets.

The max envelope (a) tends toward a rectangle as the number oftrapezoids grows. The sum envelope (b) may tend toward a bell shape or symmetric unimodal curve and some more complex curve. It does not tend toward a rectangle. The centeroid of the sum envelope in (b) must fall at or between the centeroids of the then-part trapezoids: $c_1 \le F(x) \le c_m$. The proof of the fuzzy approximation theorem depends on this fact. Centeroids ofmax envelopes can fall outside these bounds.

There are two main problems with the max combiner in $B = \overset{m}{\underset{j=1}{\vee}} a_j(x)B_j = \overset{m}{\underset{j=1}{\vee}} B_j^{'}(\text{Mamdani})$ i.e it is hard to compute centeroid and also the envelope B tends toward the rectangular or uniform (flat) shape for the net aggregate output membership function B as more and more rules fire. This increases the uncertainty of the final decision. Also the Max or Supremum nature of $B = \overset{m}{\underset{j=1}{\vee}} a_j(x)B_j = \overset{m}{\underset{j=1}{\vee}} B_j^{'}(\text{Mamdani})$ does not lead to a simple closed form solution if the fired then-part sets $B_j^{'}$ overlap. The max combiner in $B = \overset{m}{\underset{j=1}{\vee}} a_j(x)B_j = \overset{m}{\underset{j=1}{\vee}} B_j^{'}(\text{Mamdani})$ ignores the consensus in the "fired" overlap $B_j^{'} \cap B_{j+1}^{'}$. Consider the extreme case when two rules fire and have the same fired then-parts. $B_j^{'} = B_{j+1}^{'}$. The max combiner gives $B_j^{'} \vee B_{j+1}^{'} = B_j^{'}$ while the sum combiner in $B = \sum_{j=1}^{m} a_j(x)B_j\ (\text{SAM})$ gives $B_j^{'} + B_{j+1}^{'} = 2B_j^{'}$. The problem becomes more pronounced as more fired then-part sets have more overlap. Hence the more rules that fire, less the centeroidal output F(x) = centeroid (B) tends to change. The system grows less sensitive as its knowledge base grows. A limit Theorem based on BorelCantelli lemma of probability theory lies behind the max combiner's monotonic march to a rectangular pulse or the uniform probability density.

The sum combiner even though does not satisfy any known limit theorem, in practice tends toward a nearly symmetric unimodal density or bell curve, B. This observed tendency toward a bell curve as in figure. Suggests that some version ofthe central limit theorem may hold. Further research is required in this aspect. Sugeno and other Japanese fuzzy engineers have popularized the Center Of Gravity (COG) model as an adhoc scheme to "defuzzify" a fuzzy system's set output B. In this way many fuzzy engineers have used the SAM model without knowing it.

Misconceptions

Often, many misuse SAM/COG model because they vary the structure of the then-part sets B_j while they still assume that the set structure does not vary. They hand tune the sets B_j or let a learning scheme move and reshape them. Then the peaks may differ from the set centeroiods, c_j. More often the then-part set triangle, trapezoid, or bell curve are of unit height and yet adjust its width or base to reflect the rule's importance to the system output F(x). We tend to place narrow sets near equilibrium regions for more precise control and to quickly bring the system to equilibrium. But then the COG model gives more weight to the less important rules since their then-part B_j have more area or volume V_j. This can lead to needless bouts oftuning and to poor function approximation. The simple solution is either to use the full SAM system with variable set volumes V_j or to not insist that the then-part "sets" B_j have unit height and to not normalize them so to use a proper SAM system.

Generalized SAMS and the TS Model: Convex Sums of Nonlinear Systems

Additive systems include a wide range of fuzzy systems. SAM systems remain the most important class of these systems and they extend in many directions and include many models as special cases. All these systems defuzzify (usually with a centeroid) some output set B that changes with each input vector x and that arises from the sum of fired then-part sets:

$$B(x)=\sum_{i=1}^{m}B_j^{'}(x)=\sum_{j=1}^{m}a_j(x)B_j$$

These are general additive systems because the fired then-parts $B_j^{'}(x)$ depend explicitly on x. They are simple standard additive models if the input fit value $a_j(x)$ scales a fixed then-part set: $B_j^{'}(x)=a_j(x)B_j$. The fuzzy systems are generalized standard additive models if the then-part set maps each x to some non-constant output function or vector $B_j(x): B_j^{'}(x)=a_j(x)B_j(x)$. Here the then-part set B_j defines a map or operator from the input vector space to an output vector space or output function space. This extends the sum combiner (3.7) to the still-wide class of fuzzy systems that defuzzify the B(x) in:

$$B(x)=\sum_{j=1}^{m}B_j^{'}(x)=\sum_{j=1}^{m}a_j(x)B_j(x)$$

to compute F(x).

The generalized SAM still obeys some form of the SAM theorem depending on how the map or operator Bj behaves. This includes as a special linear case the socalled TSK or just TS fuzzy models of Takagi and Sugeno. All the systems compute the centeroidal output F(x) as a convex sum of linear or nonlinear systems.

Consider first the case when B_j maps each input x to a new fuzzy subset $B_j(x)\subset\Re^p$. Then we can view the then-part set function b_j as a map from a product space to real numbers or $b_j:\Re^n \text{ x } \Re^p\rightarrow[0,1]$. Each input x picks out a new then-part fuzzy set $B_j(x)$ through the restricted set function $B_j(x..):\Re^p\rightarrow[0,1]$. Then both the set volumes and centeroid depend on x and the proof of the SAM Theorem: applies otherwise with no change:

$$F(x)=\frac{\sum_{j=1}^{m}a_j(x)V_j(x)c_j(x)}{\sum_{j=1}^{m}a_j(x)V_j(x)}=\sum_{j=1}^{m}p_j(x)c_j(x)$$

This convex sum forces the user to compute each term in the SAM ratio for each input. Now suppose B_j is an arbitrary map from the input vector space, $\Re^p$ to the output vector space $\Re^p$. The map $B_j : \Re^n \to \Re^p$ is just an n-by-p matrix B_j in the linear case. In general B_j is a nonlinear operator. It l11aps each x to a new output vector $y_j = B_j(x)$. The output vector y_j depends on x but we omit this subscript for simplicity. This nonlinear operator case is a special case of the above point-to-set case when the then-part fuzzy set is just the singleton set $\{y_j\}$. This gives a unit pulse or binary set function:

$$b_j(x...) = \begin{cases} 1 & \text{if } B_j(x) = y_j \\ 0 & \text{if } B_j(x) \neq y_j \end{cases}$$

The continuous case requires us to identify the singleton set $\{y_j\}$ with a delta pulse or generalized set function:

$$b_j(x) = \delta(y - y_j) = \delta(y - y(x))$$

Then each x gives a generalized "set" $B_j(x)$ with unit volume and with a new spike or range point y_j for its centeroid, $c_j(x)$:

$$V_j(x) = \int_{\Re^p} b_j(x, y)\,dy = \int_{R^p} \delta(y - y_j)\,dy = 1$$

$$c_j(x) = \frac{\int_{\Re^p} y b_j(x, y)\,dy}{\int_{\Re^p} b_j(x, y)\,dy} = \int_{\Re^p} y\delta(y - y_j)\,dy = y_j(x)$$

Then the additive combiner in $B(x) = \sum_{i=1}^{m} B_j'(x) = \sum_{j=1}^{m} a_j(x) B_j$ further reduces to:

$$B(x) = \sum_{j=1}^{m} B_j'(x) = \sum_{j=1}^{m} a_j(x) B_j(x) = \sum_{j=1}^{m} a_j(x)\delta(y - y_j)$$

and leads to the following generalized SAM Theorem proof:

F(x) = Centroid (B(x))

$$= \frac{\int_{\Re^p} y b(x, y)\,dy}{\int_{\Re^p} b(x, y)\,dy}$$

$$= \frac{\int_{\Re^p} y \sum_{j=1}^{m} a_j(x)\delta(y - y_j)\,dy}{\int_{\Re^p} \sum_{j=1}^{m} a_j(x)\delta(y - y_j)\,dy}$$

$$= \frac{\sum_{j=1}^{m} a_j(x) \int_{\Re^p} y\delta(y - y_j)\,dy}{\sum_{j=1}^{m} a_j(x) \int_{\Re^p} \delta(y - y_j)\,dy}$$

Using the above equations,

$$= \frac{\sum_{j=1}^{m} a_j(x) y_j}{\sum_{j=1}^{m} a_j(x)} = \sum_{j=1}^{m} p_j(x) y_j$$

$$= \sum_{j=1}^{m} p_j(x) B_j(x)$$

This convex sum defines a nonlinear map $F:\Re^n \rightarrow \Re^p$ as long as m >1 or the system has two or more rules even if each then-part set B_j is a linear map or matrix. Tanaka has explored the special case of $= \sum_{j=1}^{m} p_j(x) B_j(x)$ when the operator B_j is not only a matrix but an n-by-n square matrix. Then the SAM system can define a discrete autonomous dynamical or feed back system or nonlinear difference equation:

$$x(k+1) = F(x(k)) = \sum_{j=1}^{m} p_j(x(k)) B_j(x(k))$$

The square matrix B_j can house the coefficients of a piecewise polynomial or more complex set functions b_j. The square matrix B_j can also house a separate control or forcing function $u_j(k)$.

SAM as a Function Approximator

The following theorem proves the important result that an additive fuzzy system, SAM can uniformly approximate any continuous function in a compact domain.

Theorem:

An additive fuzzy system (SAM) $F: X \rightarrow Y$ uniformly approximate $f: X \rightarrow Y$ if X is compact and *f* is continuous.

Proof:

Pick any small $\varepsilon > 0$. We must show that $|F(x) - f(x)| < \varepsilon$ for all $x \in X$, X is a compact subset of $\Re^n$ and

F (x) = Centroid(B) = Centroid $\left(\sum_{j=1}^{m} B_j^{'} \right)$ holds in an additive model.

Continuity of Ion compact X gives uniform continuity. So there is a fixed distance δ such that for all x and z in X, we have $|f(x) - f(z)| < -$ if $|x - z| < \delta$. We can construct open cubes $M_1, M_2,.., M_n$ such that to cover X and that have ordered overlap in their n coordinates so that each cube corner lies at the mid point c_j of its neighbors M_j. Pick then-part fuzzy sets B_j centered at $f(c_j)$. Thus the centeroid of B_j is $f(c_j)$.

Pick $u \in X$. Then by construction u lies in at most 2^n overlapping open cubes M_j. Pick any w in the same set of cubes. Suppose $u \in M_j$ and $w \in M_k$. Then for all $v \in M_j \cap M_k$, we have $|u-v| < \delta$ and $|v-w| < \delta$. Uniform continuity implies that

$$|f(u)-f(w)| \le |f(u)-f(v)| + |f(v)-f(w)| < \frac{\varepsilon}{2}.$$

So for cube centers c_j and c_k, we have $|f(c_j)-f(c_k)| < \frac{\varepsilon}{2}$.

Pick $x \in X$. Then x too lies in at most 2^n open cubes with centers c_j and such that $|f(c_j)-f(x)| < \frac{\varepsilon}{2}$. Along the kth coordinate of the range space $\Re^p$, the kth component of the additive system centeroid F(x) lies on or between the kth components of the centeroids of the then-part B_j sets. So $|F(x)-f(c_j)| < \frac{\varepsilon}{2}$ holds since $|f(c_j)-f(c_k)| < \frac{\varepsilon}{2}$ holds for all $f(c_j)$. Then $|F(x)-f(x)| \le |F(x)-f(c_j)| + |f(c_j)-f(x)| < \frac{\varepsilon}{2}+\frac{\varepsilon}{2} = \varepsilon$

Exact Scalar Representation using SAM

For the exact fuzzy modeling of a known scalar function, the following Theorem: uses the bounds of the scalar function for normalization and builds membership functions, absorbing the function itself in the membership functions. The following theorem is used and extended for the exact TS fuzzy dynamic modeling of a class of nonlinear dynamic system.

Theorem:

A SAM $F: \Re \to \Re$ with just two rules of the form

"If X is M_1 then Y is B1"

"If X is (not – M_1) then Y is B_2"

can represent a bounded non-constant function $f: \Re \to \Re$ in the sense that F(x) = f(x) for all $x \in \Re$

Proof:

Bounded-ness let us define the lower and upper bounds α = min (f), β = max (f). Then-part sets B_1 and B_2 of any shape with centeroids at α and β with finite but constant volumes (areas) : $0 < V_1 = V_2 < \infty$. So C1 = α and C_2 = β. B_1 and B_2 might be rectangles or triangles of unit area. Define the set function of if-part set M_1 as the ratio.

$$M_1(x) = \frac{\beta - f(x)}{\beta - \alpha}$$

then $0 \le M_1(x) \le 1$ holds for all x and so M_1 is a fuzzy set. Define the second then part set M_2 as the complement of M_1 : $M_2(x) = 1 - M_1(x)$. Then $M_1(x) = \frac{\beta - f(x)}{\beta - \alpha}$ and Theorem give the following result:

$$F(x) = \frac{\sum_{j=1}^{2} M_j(x) V_j c_j}{\sum_{j=1}^{2} M_j(x) V_j}$$

$$= \frac{M_1(x)\alpha + (1 - M_1(x))\beta}{M_1(x) + 1 - M_1(x)}$$

$$= M_1(x)\alpha + (1 - M_1(x))\beta$$
$$= M_1(x)(\alpha - \beta) + \beta$$
$$= \left[\frac{\beta - f(x)}{\beta - \alpha}\right](\alpha - \beta) + \beta$$
$$= f(x) - \beta + \beta = f(x)$$

Example:

Consider the bounded function f(x) = sin x. Pick the two then-part sets B_1 and B_2 as unit area rectangles centered at –1 and 1:

$$C_1 = \alpha = -1 \text{ and } C_2 = \beta = 1 = V_1 = V_2. \text{Then}$$

$$M_1(x) = \frac{1 - \sin x}{1 - (-1)} = \frac{1 - \sin x}{2} \text{ and}$$

$$M_2(x) = \frac{1 + \sin x}{2}$$

Then from Theorem:

$$F(x) = \frac{\left(\frac{1-\sin x}{2}\right)(-1) + \left(\frac{1+\sin x}{2}\right)(1)}{\frac{1-\sin x}{2} + \frac{1+\sin x}{2}}$$

$$= \frac{\sin x - 1 + 1 + \sin x}{1 - \sin x + 1 + \sin x} = \frac{2\sin x}{2} = \sin x$$

Such fuzzy representations reduce the exponential rule complexity of blind function approximation to linear (or quadratic) complexity but at the cost of a complete knowledge of f. Future research may find algorithms that reduce the SAM rule complexity from exponential to polynomial as the system learns more of the structure of f.

MAMDANI FUZZY MODEL

The Mamdani fuzzy inference system was proposed as the first attempt to control a steam engine and boiler combination by a set of linguistic control rules obtained from experienced human operators. figure is an illustration of how a two-rule Mamdani fuzzy inference system derives the overall output z when subjected to two crisp inputs x and y.

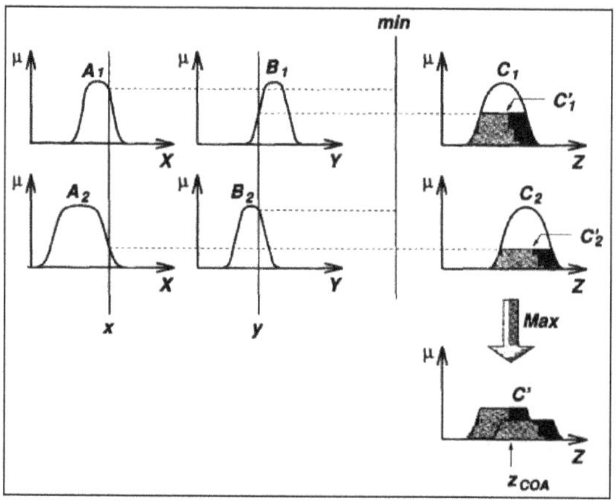

The Mamdani fuzzy inference system using min and max for T-norm and T-conorm operators, respectively.

If we adopt max and algebraic product as our choice for the T-norm and T-conorm operators, respectively, and use max-product composition instead of the original max-min composition, then the resulting fuzzy reasoning is shown in figure, where the inferred output of each rule is a fuzzy set scaled down by its firing strength via algebraic product. Although this type of fuzzy reasoning was not employed in Mamdani's original paper, it has often been used in the literature. Other variations are possible if we use different T-norm and T-conorm operators.

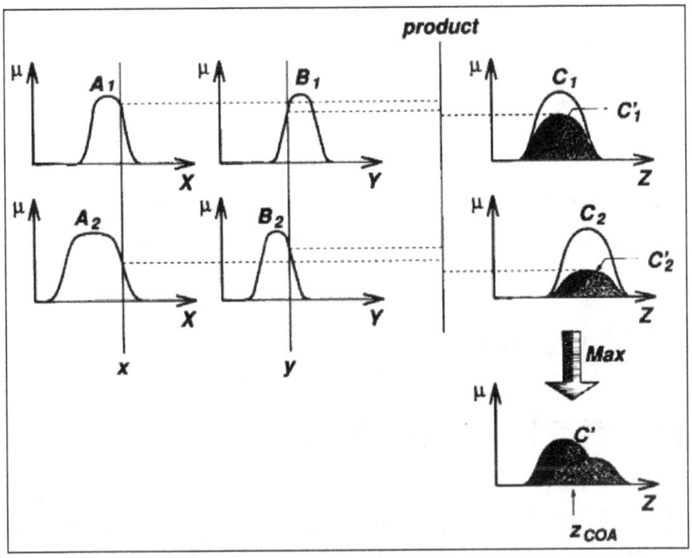

The Mamdani fuzzy inference system using product and max for T-norm and T-conorm operators, respectively.

In Mamdani's application, two fuzzy inference systems were used as two controllers to generate the heat input to the boiler and throttle opening of the engine cylinder, respectively, to regulate the steam pressure in the boiler and the speed of the engine. Since the plant takes only crisp values as inputs, we have to use a defuzzifier to convert a fuzzy set to a crisp value.

Defuzzification

Defuzzification refers to the way a crisp value is extracted from a fuzzy set as a representative value. In general, there are five methods for defuzzifying a fuzzy set A of a universe of discourse Z, as shown in figure. (Here the fuzzy set A is usually represented by an aggregated output MF, such as C' in figures).

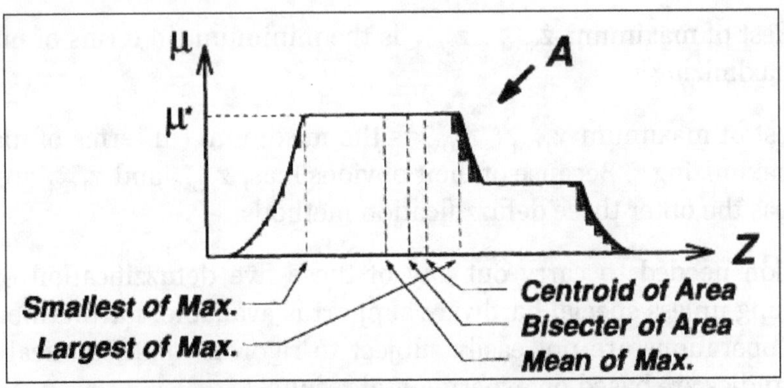

Various defuzzification schemes for obtaining a crisp output.

A brief explanation of each defuzzification strategy are shown as follows:

- Centroid of area z_{COA} :

$$z_{COA}=\frac{\int_Z \mu_A(z)dz}{\int_Z \mu_A(z)dz},$$

where $\mu_A(z)$ is the aggregationed output MF. This is the most widely adopted defuzzification strategy, which is reminiscent of the calculation of expected values of probability distributions.

- Bisector of area z_{BOA} : z_{BOA} satisfies

$$\int_\alpha^{z_{BOA}} \mu_A(z)dz = \int_{z_{BOA}}^{\beta} \mu_A(z)dz,$$

where $\alpha = \min\{z \mid z \in Z\}$ and $\beta = \max\{z \mid z \in Z\}$. That is, the vertical line $z = z_{BOA}$ partitions the region between $z = \alpha$, $z = \beta$, $y = 0$ and $y = \mu_A(z)$ into two regions with the same area.

- Mean of maximum z_{MOM} : z_{MOM} is the average of the maximizing z at which the MF reach a maximum μ^*. In symbols:

$$z_{MOM}=\frac{\int_{Z'} z\,dz}{\int_{Z'} dz},$$

where $Z' = \{z \mid \mu_A(z) = \mu^*\}$. In particular, if $\mu_A(z)$ has a single maximum at $z = z^*$, then $z_{MOM} = z^*$. Moreover, if $\mu_A(z)$ reaches its maximum whenever $z \in [z_{left}, z_{right}]$ (This is the case in figure), then $zMOM = (z_{left} + z_{right})/2$. The mean of maximum is the defuzzification strategy employed in Mamdani's fuzzy logic controllers.

- Smallest of maximum z_{SOM} : z_{SOM} is the minimum (in terms of magnitude) of the maximizing z.

- Largest of maximum z_{LOM} : z_{LOM} is the maximum (in terms of magnitude) of the maximizing z. Because of their obvious bias, z_{SOM} and z_{LOM} are not used as often as the other three defuzzification methods.

The calculation needed to carry out any of these five defuzzification operations is time-consuming unless special hardware support is available. Furthermore, these defuzzification operations are not easily subject to rigorous mathematical analysis, so most of the studies are based on experimental results.

A Mamdani Type Fuzzy Logic Controller

The database of a rule-based system may contain imprecisions which appear in the description of the rules given by the expert. Because such an inference can not be made by the methods which use classical two valued logic or many valued logic, Zadeh in and Mamdani in suggested an inference rule called "compositional rule of inference". Using this inference rule, several methods for fuzzy reasoning were proposed. Zadeh extends the traditional Modus Ponens rule in order to work with fuzzy sets, obtaining the Generalized Modus Ponens (GMP) rule.

An important part of fuzzy reasoning is represented by Fuzzy Logic Control (FLC), derived from control theory based on mathematical models of the open-loop process to be controlled. Fuzzy Logic Control has been successfully applied to a wide variety of practical problems: control of warm water, robot, heat exchange, traffic junction, cement kiln, automobile speed, automotive engineering, model car parking and turning, power system and nuclear reactor, on-line shopping, washing machines, etc.

It points out that fuzzy control has been effectively used in the context of complex ill-defined processes, especially those that can be controlled by a skilled human operator without the knowledge of their underlying dynamics. In this sense, neural and adaptive fuzzy systems has been compared to classical control methods by B. Kosko in. There, it is remarked that they are model-free estimators, i.e., they estimate a function without requiring a mathematical description of how the output functionally depends on the input; they learn from samples. However, some people criticized fuzzy control because the very fundamental question "Why does a fuzzy rule-based system have such good performance for a wide variety of practical problems?" remained unanswered. A first approach to answer this fundamental question in a quantitative way was presented by Wang in where he proved that a particular class of FLC systems are universal approximators: they are capable of approximating any real continuous function on a compact set to arbitrary accuracy. This class is characterized by:

- Gaussian membership functions,
- Product fuzzy conjunction,
- Product fuzzy implication,
- Center of area defuzzification.

Other approaches are due to Buckley. He has proved that a modification of Sugeno type fuzzy controllers gives universal approximators. Although both results are very important, many real fuzzy logic controllers do not belong to these classes, because other membership functions are used, other inference mechanisms are applied or other type of rules are used.

The most popular FLC systems are: Mamdani, Tsukamoto, Sugeno and Larsen which work with crisp data as inputs. An extension of the Mamdani model in order to work with interval inputs is presented in, where the fuzzy sets are represented by triangular fuzzy numbers and the firing level of the conclusion is computed as the product of firing levels from the antecedent. Other extensions and applications of the standard FLC systems were proposed.

The necessity to extend the fuzzy controllers to work with intervals or linguistic values as inputs is given by many applications where precise values of the input data no interest or are difficult to estimate. For example, in shopping applications the buyer is interested, rather, in a product that is priced within certain limits or does not exceed a given value. In other cases, the input values are much easier to express in fuzzy manner, for example, in the problem of controlling the washing time using fuzzy logic control the degree of dirt for the object to be washed is easily expressed by a linguistic value. These examples will be used to show the working of the model proposed in order to expand the Mamdani fuzzy logic controller. A FLC system has the following characteristics:

- The linguistic terms (or values) are represented by trapezoidal fuzzy numbers.
- Various implication operators are used to represent the rules.
- The crisp control action of a rule is computed using Middle-of-Maxima method.
- The overall crisp control action of an implication is computed by discrete Center-of-Gravity.
- The overall crisp control action of the system is computed using an OWA (Ordered Weighted Averaging) operator.

Preliminaries

Let U be a collection of objects denoted generically by {u}, which could be discrete or continuous. U is called the universe of discourse and u represents the generic element of U.

A fuzzy set F in the universe of discourse U is characterized by its membership function $\mu_F : U \to [0,\ 1]$. The fuzzy set may be represented as a set of ordered pairs of a generic element u and its grade of membership function: $F = \{(u,\quad F(u)) / u \in U\}$.

A fuzzy number F in a continuous universe U, e. g., a real line, is a fuzzy set F in U which is normal and convex, i. e.,

$$\max_{\mu \in U} \mu_F(u) = 1 \qquad (\text{normal})$$

$$\mu_F(\lambda u_1 + (1\ \lambda)u_2) \geq \min\{\mu_F(u_1),\ \mu_F(u_2)\},\ u_1,\ u_2\ \in,\ U,\ \lambda \in,\ [0,\ 1]\ (convex)$$

Such a number N $=\left(\underline{m},\overline{m},\alpha,\beta\right)$ is defined as:

$$\mu_N(x)=\begin{cases} 0 \ for \ x<\underline{m}-\alpha \\ \dfrac{x-\underline{m}+\alpha}{\alpha} \ for \ x\in\left[\underline{m}-\alpha,\underline{m}\right] \\ 1 \ for \ x\in\left[\underline{m},\overline{m}\right] \\ \dfrac{\overline{m}+\beta-x}{\beta} \ for \ x\in\left[\overline{m},\overline{m}+\beta\right] \\ 0 \ for \ x>\overline{m}+\beta \end{cases}$$

Will be used fuzzy sets to represent linguistic variables. A linguistic variable can be regarded either as a variable whose value is a fuzzy number or as a variable whose values are defined in linguistic terms.

A linguistic variable V is characterized by: its name x, an universe U, a term set T(x), a syntactic rule G for generating names of values of x, and a set of semantic rule M for associating with each value its meaning.

For example, if speed of a car is interpreted as a linguistic variable, then its term set could be T(x) = {slow, moderate, fast, very slow, more or less fast} where each term is characterized by a fuzzy set in an universe of discourse U = [0, 100]. We might interpret: slow as "a speed below about 40 mph", moderate as "speed close to 55 mph", fast as "a speed about 70 mph".

A function $T:[0,\ 1]^2\rightarrow[0,\ 1]$ is a t-norm iff it is commutative, associative, non-decreasing and $T(x,\ 1)=x\forall x\in[0,\ 1]$.

The most important t-norms are:

- Minimum: $T_m(x,\ y)=\min\{x,\ y\}$
- Lukasiewicz: $T_L(x,\ y)=\max\{0,\ x\ +\ y\ -\ 1\}$
- Probabilistic (or Product): $T_P(x,\ y)=xy$
- Weak: $T_W(x,\ y)\ =\begin{cases}\min\{x,y\}\ if & \max\{x,y\}=1\\ 0 & otherwise\end{cases}$

A function $S:[0,\ 1]^2\rightarrow[0,\ 1]$ is a t-conorm iff it is commutative, associative, non-decreasing and $S(x,\ 0)=x\forall x\in[0,\ 1]$.

The basic t-conorms are:

- Maximum: $S_m(x,\ y)=\max\{x,\ y\}$

- Lukasiewicz: $S_L(x, y) = \min\{1, x + y\}$
- Probabilistic (or Product): $S_P(x, y) = x + y - xy$
- Strong: $S_S(x, y) = \begin{cases} \max\{x, y\} \text{ if } & \min\{x, y\} = 1 \\ 0 & \text{otherwise} \end{cases}$

The t-norms are used to compute the firing levels of the rules or as aggregation operators and the t-conorms are used as aggregation operators. The rules are represented by fuzzy implications. Let X and Y be two variables whose domains are U and V, respectively. A causal link from X to Y is represented as a conditional possibility distribution $\pi Y/X$ which restricts the possible values of Y for a given value of X. For the rule

IF X is A THEN Y is B

we have,

$$\forall u \in U, \forall v \in V, \pi_{Y/X}(v, u) = \mu_A(u) \rightarrow \mu_B(v)$$

Where $\rightarrow$ is an implication operator and μ_A and μ_B are the membership functions of the fuzzy sets A and B, respectively.

An implication is a function $I : [0, 1]^2 \rightarrow [0, 1]$ satisfying the following conditions for all $x, y, z \in [0, 1]$:

- I1: If x ≤ z then I(x, y) ≥ I(z, y).
- I2: If y ≤ z then I(x, y) ≤ I(x, z).
- I3: I(0, y) = 1 (falsity implies anything).
- I4: I(x, 1) = 1 (anything implies tautology).
- I5: I(1, 0) = 0 (Booleanity).

The following properties could be important in some applications:

- I6: I(1, x) = x (tautology cannot justify anything).
- I7: I(x, I(y, z)) = I(y, I(x, z)) (exchange principle).
- I8: x ≤ y if and only if I(x, y) = 1 (implication defines ordering).
- I9: I(x, 0) = N(x) is a strong negation.
- I10: I(x, y) ≥ y.
- I11: I(x, x) = 1 (identity principle).

- I12: I(x, y) = I(N(y), N(x)), where N is a strong negation.
- I13: I is a continuous function.

The most important implications are:

- Willmott: $I_W(x, y) = \max\{1-x, \ \min\{x, y\}\}$.
- Mamdani: $I_M(x, y) = \min\{x, y\}$.
- Rescher-Gaines: $I_{RG}(x, y) = \begin{cases} 1 \ \mathit{if} & x \le y \\ 0 & \mathit{otherwise} \end{cases}$.
- Kleene-Dienes: $I_{KD}(x, y) = \max\{1 - x, y\}$.
- Brouwer-Gödel: $I_{BG}(x, y) = \begin{cases} 1 \ \mathit{if} & x \le y \\ y & \mathit{otherwise} \end{cases}$.
- Goguen: $I_G(x, y) = \begin{cases} 1 \mathit{if} & x \le y \\ \dfrac{y}{x} & \mathit{otherwise} \end{cases}$.
- Lukasiewicz: $I_L(x, y) = \min\{1 - x + y, 1\}$
- Fodor: $I_F(x, y) = \begin{cases} 1 & \mathit{if} \ x \le y \\ \max\{1-x, y\} & \mathit{otherwise} \end{cases}$.
- Reichenbach: $I_R(x, y) = 1 - x + xy$.

An n-ary fuzzy relation is a fuzzy set in $U_1 \times U_2 \times \cdots \times U_n$ expressed as:

$$R_{U_1} \times .... \times u_n \{((u_1, ..., u_n), \mu_R(u_1, ..., u_n)) / (u_1, ... u_n) \in U_1 \times .... \times U_n\}.$$

If R and S are fuzzy relations in U × V and V ×W, respectively, then the sup-star composition of R and S is a fuzzy relation denoted by $R \cdot S$ and defined by:

$$R \cdot S = \{[(u, w), \sup_v(\mu_R(u, v) * \mu_S(v, m))] / u \in U, v \in V, w \in W\}$$

where * can be any operator in the class of t-norms.

Fuzzy implication inference is based on the compositional rule of inference for approximate reasoning suggested by Zadeh.

If R is a fuzzy relation on U × V and x is a fuzzy set in U then the "sup-star compositional rule of inference" asserts that the fuzzy set y in V induced by x is given by:

$$y = x \cdot R$$

where x · R is the sup-star composition of x and R.

If the star represents the minimum operator then this definition reduces to Zadeh's compositional rule of inference.

The process of information aggregation appears in many applications related to the development of intelligent systems: fuzzy logic controllers, neural networks, vision systems, expert systems, multi-criteria decision aids. In Yager introduced an aggregation technique based on OWA operators.

An OWA operator of dimension n is a mapping $F : R^n \to R$ that has an associated n vector $w = (w_1, w_2, ..., w_n)^t$ such as:

$$w_i \in [0, 1], 1 \le i \le n, \sum_{i=1}^{n} w_i = 1.$$

The aggregation operator of the values $\{a_1, a_2, ..., a_n\}$ is:

$$F(a_1, a_2, ..., a_n) = \sum_{j=1}^{n} w_j b_j$$

where b_j is the j-th largest element from $\{a_1, a_2, ..., a_n\}$.

It is sufficiently to work with rules with a single conclusion because a rule with multiple consequent can be treated as a set of such rules.

Standard Fuzzy Logic Controllers

Structure of a Fuzzy Logic Controller

The seminal work by L.A. Zadeh on fuzzy algorithms introduced the idea of formulating the control algorithm by logical rules. An FLC consists of a set of rules of the form:

IF (a set of conditions are satisfied) THEN (a set of consequences can be in ferred).

Since the antecedents and the consequents of these IF-THEN rules are associated with fuzzy concepts (linguistic terms), they are often called fuzzy conditional statements. In FLC terminology, a fuzzy control rule is a fuzzy conditional statement in which the antecedent is a condition in its application domain and the consequent is a control action for the system under control. The inputs of fuzzy rule-based systems should be given by fuzzy sets, and therefore, we have to fuzzify the crisp inputs. Furthermore, the output of a fuzzy system is always a fuzzy set, and therefore to get crisp value we have to defuzzify it. Fuzzy logic control systems usually consist of four major parts: Fuzzification interface, Fuzzy rule base, Fuzzy inference engine and Defuzzification interface, as is presented in the figure.

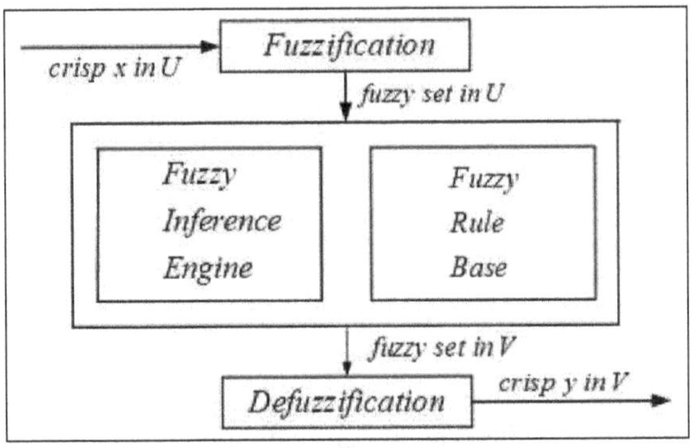

Fuzzy Logic Controller.

The four components of a FLC are explained in the following:

The fuzzification interface involves the functions:

- Measures the values of inputs variables,
- Performs a scale mappings that transfers the range of values of inputs variables into corresponding universes of discourse,
- Performs the function of fuzzyfication that converts input data into suitable linguistic values which may be viewed as label of fuzzy sets.

The rule base comprises a knowledge of the application domain and the attendant control goals. It consists of a "data base" and a "linguistic (fuzzy) control rule base":

- The data base provides necessary definitions which are used to define linguistic control rules and fuzzy data manipulation in a FLC.
- The rule base characterizes the control goals and the control policy of the domain experts by means of a set of linguistic control rules.

The fuzzy inference engine is the kernel of a FLC; it has the capability of simulating human decision-making based of fuzzy concepts and of inferring fuzzy control actions employing fuzzy implication and the rules of inference in fuzzy logic.

The defuzzification interface performs the following functions:

- A scale mapping, which converts the range of values of output variables into corresponding universes of discourse.
- Defuzzification, which yields a non fuzzy control action from an inferred fuzzy control action.

A fuzzification operator has the effect of transforming crisp data into fuzzy sets. In most of the cases fuzzy singletons are used as fuzzifiers.

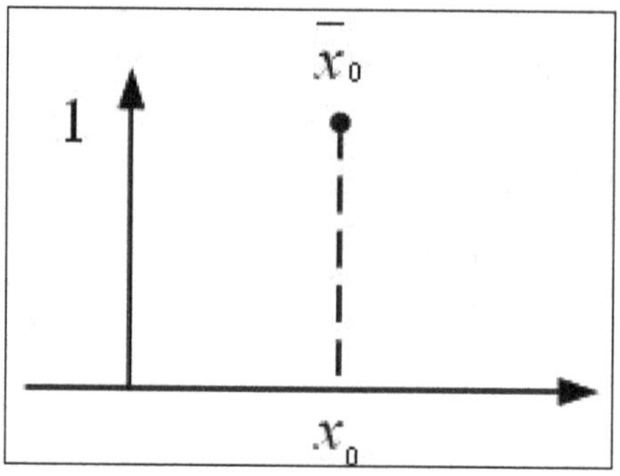

Fuzzy singleton as fuzzifier.

In other words,

$$fuzzifier(x_0) = \bar{x}0,$$

$$\mu_{\bar{x}0}(x) = \begin{cases} 1 & for \;\; x = x_0 \\ 0 & x \neq x_0 \end{cases}$$

where x_0 is a crisp input value from a process.

The procedure used by Fuzzy Inference Engine in order to obtain a fuzzy output consists of the following steps:

- Find the firing level of each rule,
- Find the output of each rule,
- Aggregate the individual rules outputs in order to obtain the overall system output.

The fuzzy control action C inferred from the fuzzy control system is transformed into a crisp control action:

$$z_0 = defuzzifier(C)$$

where de fuzzifier is a defuzzification operator. The most used defuzzification operators, for a discrete fuzzy set C having the universe of discourse V, are:

- Center-of-Gravity:

$$z_0 = \frac{\sum_{j=1}^{N} z_j \mu_C(z_j)}{\sum_{j=1}^{N} \mu_C(z_j)}$$

- Middle-of-Maxima: the defuzzified value is defined as mean of all values of the universe of discourse, having maximal membership grades:

$$z_0 = \frac{1}{N1}\sum_{j=1}^{N1} z_j,\ N1 \leq N$$

- Max-Criterion: this method chooses an arbitrary value, from the set of maximizing elements of C, i. e.

$$z_0 \in, \{z / \mu_C(z) = \max_{v \in V} \mu_C(v)\}$$

where $Z = \{z_1, \ldots, z_N\}$ is a set of elements from the universe V.

Because several linguistic variables are involved in the antecedents and the conclusions of a rule, the fuzzy system is of the type multi–input–multi–output. Further, the working with a FLC for the case of a two-input-single-output system is explained. Such a system consists of a set of rules:

R_1 : IF x is A_1 AND y is B_1 THEN z is C_1

R_2 : IF x is A_2 AND y is B_2 THEN z is C_2

R_n : IF x is A_n AND y is B_n THEN z is C_n

and a set of inputs,

fact : x is x_0 AND y is y_0

where x and y are the process state variables, z is the control variable, A_i, B_i and C_i are linguistic values of the linguistic variables x, y and z in the universes of discourse U, V and W, respectively. Our task is to find a crisp control action z_0 from the fuzzy rule base and from the actual crisp inputs xo and y_0. A fuzzy control rule:

R_i : IF x is A_i AND y is B_i THEN z is C_i

is implemented by a fuzzy implication I_i and is defined as:

$$\mu_{I_i}(u,\ v, w) = [\mu_{A_i}(u)\ AND\ \mu_{B_i}(v)] \rightarrow \mu_{C_i}(w) = T(\mu_{A_i}(u),\ \mu_{B_i}(v)) \rightarrow \mu_{C_i}(w)$$

where T is a t-norm used to model the logical connective AND. To infer the consequence "z is C" from the set of rules and the facts, usually the compositional rule of inference is applied; it gives,

$$consequence\ =\ Agg\{fact \cdot R_1,\ \ldots,\ fact \cdot R_n\}.$$

That is,

$$\mu_C = Agg\left\{T\left(\mu_{\bar{x0}}, \mu_{\bar{y}_0}\right)\cdot R_1,\ldots,T\left(\mu_{\bar{x0}}, \mu_{\bar{y}_0}\right)\cdot R_n\right\}.$$

Taking into account that $\mu_{\bar{x0}}$ (u) = 0 for u ≠ x_o and $\mu_{\bar{y}_0}$ (v) = 0 for v ≠ y_o, the membership function of C is given by:

$$\mu_C(w) = Agg\left\{T\left(\mu_{A_1}(x_0),\mu_{B_1}(y_0)\right)\rightarrow \mu_{C_1}(w),\ldots.,T\left(\mu_{A_n}(x_0),\mu_{B_n}(y_0)\right)\rightarrow \mu_{C_n}(w)\right\}$$

for all w ∈ W.

The procedure used for obtaining the fuzzy output from a FLC system is:

- The firing level of the i-th rule is determined by:

  $$T(\mu_{A_i}(x_0),\ \mu_{B_i}(y_0))$$

- The output $C_i^{'}$ of the i-th rule is given by:

  $$\mu_{C_i^{'}}(w) = T\left(\mu_{A_i}(x_0),\mu_{B_i}(y_0)\right)\rightarrow \mu_{C_i}(w)$$

- The overall system output, C, is obtained from the individual rule outputs, by aggregation operation:

  $$\mu_C(w) = Agg\left\{\mu_{C_1^{'}}(w),\ldots,\mu C_n^{'}(w)\right\}$$

for all w ∈ W.

Mamdani Fuzzy Logic Controller

The most commonly used fuzzy inference technique is the so-called Mamdani method which was proposed, by Mamdani and Assilian, as the very first attempt to control a steam engine and boiler combination by synthesizing a set of linguistic control rules obtained from experienced human operators. Their work was inspired by an equally influential publication by Zadeh. Interest in fuzzy control has continued ever since, and the literature on the subject has grown rapidly. A survey of the field with fairly extensive references may be found in or, more recently, in. In Mamdani's model the fuzzy implication is modeled by Mamdani's minimum operator, the conjunction operator is min, the t-norm from compositional rule is min and for the aggregation of the rules the max operator is used. In order to explain the working with this model of FLC will be considered the example from where a simple two-input one-output problem that includes three rules is examined:

Rule1 : IF x is A_3 OR y is B_1 THEN z is C_1

Rule2 : IF x is A_2 AND y is B_2 THEN z is C_2

Rule3 : IF x is A_1 THEN z is C_3.

Step 1: Fuzzification.

The first step is to take the crisp inputs, x_0 and y_0, and determine the degree to which these inputs belong to each of the appropriate fuzzy sets. According to figure one obtains

$$\mu_{A_1}(x_0)=0.5,\ \mu_{A_2}(x_0)=0.2,\ \mu_{B_1}(y_0)=0.1,\ \mu_{B_2}\ (y_0)=0.7$$

Step 2: Rules evaluation.

The fuzzified inputs are applied to the antecedents of the fuzzy rules. If a given fuzzy rule has multiple antecedents, the fuzzy operator (AND or OR) is used to obtain a single number that represents the result of the antecedent evaluation. To evaluate the disjunction of the rule antecedents, one uses the OR fuzzy operation. Typically, the classical fuzzy operation union is used:

$$\mu_{A\cup B}(x)=\max\{\mu_A(x),\ \mu_B(x)\}.$$

Similarly, in order to evaluate the conjunction of the rule antecedents, the AND fuzzy operation intersection is applied:

$$\mu_{A\cup B}(x)=\min\{\mu_A(x),\ \mu_B(x)\}.$$

Now the result of the antecedent evaluation can be applied to the membership function of the consequent. The most common method is to cut the consequent membership function at the level of the antecedent truth; this method is called clipping. Because top of the membership function is sliced, the clipped fuzzy set loses some information. However, clipping is preferred because it involves less complex and generates an aggregated output surface that is easier to defuzzify. Another method, named scaling, offers a better approach for preserving the original shape of the fuzzy set: the original membership function of the rule consequent is adjusted by multiplying all its membership degrees by the truth value of the rule antecedent.

Step 3: Aggregation of the rule outputs.

The membership functions of all rule consequents previously clipped or scaled are combined into a single fuzzy set.

Step 4: Defuzzification.

The most popular defuzzification method is the centroid technique. It finds a point representing the center of gravity (COG) of the aggregated fuzzy set A, on the interval [a, b]. A reasonable estimate can be obtained by calculating it over a sample of points. According to figure in our case results:

$$\text{COG} = \frac{(0+10+20) \times 0.1+(30+40+50+60) \times 0.2+(70+80+90+100) \times 0.5}{0.1 + 0.1 + 0.1 + 0.2 + 0.2 + 0.2 + 0.2 + 0.5 + 0.5 + 0.5 + 0.5} = 67.4$$

Universal Approximators

Using the Stone-Weierstrass theorem, Wang in showed that fuzzy logic control systems of the form:

$$R_i : \text{IF x is } A_i \text{ AND y is } B_i \text{ THEN z is } C_i,\ i = 1, \ldots, n$$

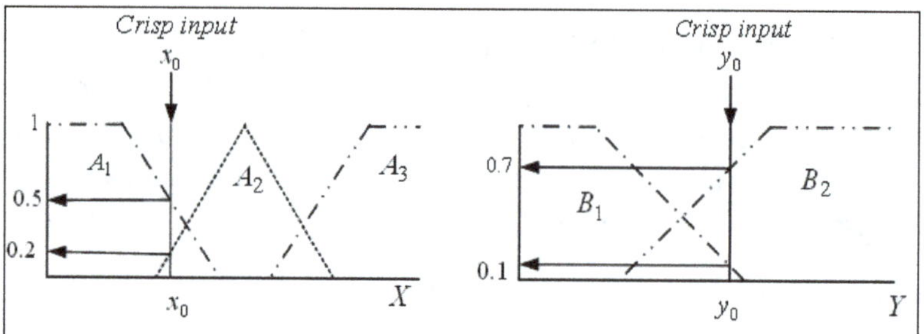

(a) Fuzzification

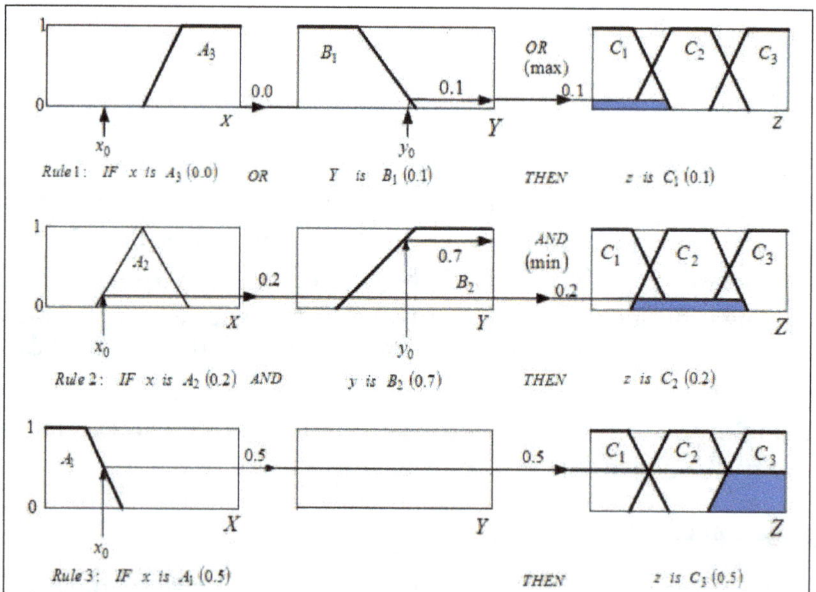

(b) Rules evaluation

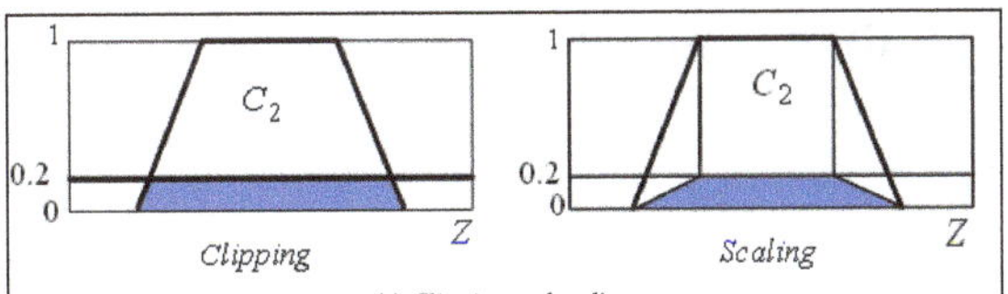

(c) Clipping and scaling

Mamdani fuzzy logic controller.

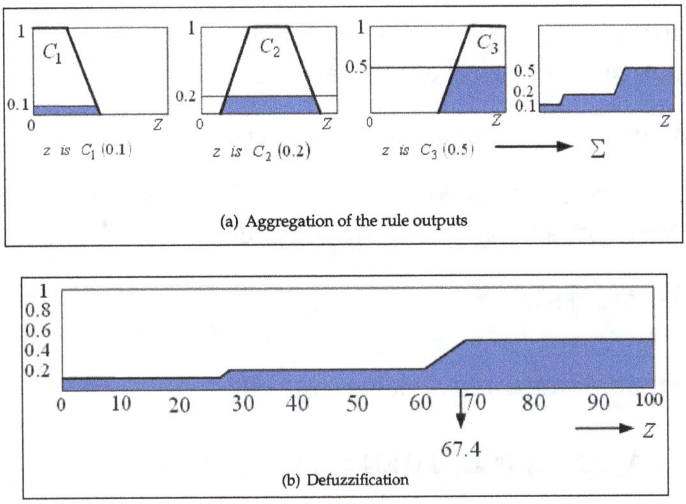

(a) Aggregation of the rule outputs

(b) Defuzzification

Mamdani fuzzy logic controller.

- Gaussian membership functions:

$$\mu_A(x) = \exp\left[-\frac{1}{2}\left(\frac{x - x_0}{\sigma}\right)^2\right]$$

where x_o is the position of the peak relative to the universe and σ is the standard deviation.

- Singleton fuzzifier:

$$fuzzifier(x) = \bar{x}$$

- Fuzzy product conjunction:

$$\mu_{A_i}(u)\ AND\ \mu_{B_i}(v) = \mu_{A_i}(u)\mu_{B_i}(v)$$

Minimum-norm fuzzy implication:

$$[\mu A_i(u)\ AND\ \mu B_i(v)] \rightarrow \mu c_i(w) = \min\{\mu A_i(u), \mu B_i(v), \mu C_i(w)\}$$

- Larsen (fuzzy product) implication:

$$[\mu_{A_i}(u)\ AND\ \mu_{B_i}(v)] \rightarrow \mu_{C_i}(w) = \mu_{A_i}(u)\mu_{B_i}(v)\ \mu_{C_i}(w)$$

- Centroid deffuzification method:

$$z = \frac{\sum_{i=1}^{n} c_i \mu_{A_i}(x)\mu_{B_i}(y)}{\sum_{i=1}^{n} \mu_{A_i}(x)\mu_{B_i}(y)}$$

where c_i is the center of C_i, are universal approximators, i.e. they can approximate any continuous function on a compact set to an arbitrary accuracy.

More generally, Wang proved the following theorem:

Theorem: For a given real-valued continuous function g on the compact set U and arbitrary $\varepsilon > 0$, there exists a fuzzy logic control system with output function f such that:

$$\sup_{x \in U} \left\| g(x) - f(x) \right\| \leq \varepsilon.$$

Castro in showed that Mamdani fuzzy logic controllers:

R_i : IF x is A_i AND y is B_i THEN z is C_i, i = 1, ..., n

With

- Symmetric triangular membership functions:

$$\mu_A(x) = \begin{cases} 1 - \frac{x-a}{\alpha} \, if & |x-a| \leq \alpha \\ 0 & otherwise \end{cases}$$

- Singleton fuzzifier:

$$fuzzifier(x_0) = \bar{x}_0 \hat{}$$

- Minimum norm fuzzy conjunction:

$$\mu_{A_i}(u) \; AND \; \mu_{B_i}(v) = \min\{\mu_{A_i}(u), \; \mu_{B_i}(v)\}$$

- Maximum t-conorm rule aggregation:

$$Agg\{R_1, R_2, ..., R_n\} = \max\{R_1, R_2, ..., R_n\}$$

- Centroid defuzzification method:

$$z = \frac{\sum_{i=1}^{n} c_i \min\{\mu_{A_i}(x), \mu_{B_i}(y)\}}{\sum_{i=1}^{n} \min\{\mu_{A_i}(x), \mu_{B_i}(y)\}}$$

where c_i is the center of C_i, are universal approximators.

More generally, Castro studied the following problem:

Given a type of FLC, (i.e. a fuzzification method, a fuzzy inference method, a defuzzification method, and a class of fuzzy rules, are fixed), an arbitrary continuous real valued function f on a compact $U \subset R^n$, and a certain $\varepsilon > 0$, is it possible to find a set of fuzzy rules such that the associated fuzzy controller approximates f to level ε?

The main result obtained by Castro is that the approximation is possible for almost any type of fuzzy logic controller.

Mamdani FLC with Different Inputs and Implications

Further, the standard Mamdani FLC system will be extended to work as inputs with crisp data, intervals and linguistic terms and with various implications to represent the rules. A rule is characterized by:

- A set of linguistic variables A, having as domain an interval $I_A = [a_A, bA]$.
- n_A linguistic values $A_1, A_2, \dots, A_{n_A}$ for each linguistic variable A.
- The membership function for each value A_i, denoted as $\mu_A^0(x)$ where $i \in \{1, 2, \dots, n_A\}$ and $x \in I_A$.

Fuzzification

A fuzzification operator transforms a crisp data or an interval into a fuzzy set. For instance, $x_0 \in \mathrm{U}$ is fuzzified into $\bar{x}_0$ according with the relation:

$$\mu_{\bar{x}_0}(x) = \begin{cases} 1 & \textit{if } x = x_0 \\ 0 & \textit{otherwise} \end{cases}$$

and an interval input [a, b] is fuzzified into:

$$\mu_{[a,b]}(x) = \begin{cases} 1 & \textit{if } x \in [a,b] \\ 0 & \textit{otherwise} \end{cases}$$

Firing Levels

The firing level of a linguistic variable A_i, which appears in the premise of a rule, depends of the input data.

- For a crisp value x_0 it is $\mu^0_{A_i}(x_0)$.

If the input is an interval or a linguistic term then the firing level can be computed in various forms.

- Based on "intersection"
- For an input interval [a, b] it is given by:

$$\mu_{A_i} = \max\left\{\min\left\{\mu^0_{A_i}(x), \mu_{[a,b]}(x)\right\} \middle| x \in [a,b]\right\}.$$

- For a linguistic input value $A_i^{'}$ it is:

$$\mu_{A_i} = \max\left\{\min\left\{\mu^0_{A_i}(\mathrm{x}),\ \mu_{A_i^{'}}(\mathrm{x})\right\} \middle| x \in I_A\right\}.$$

- Based on "areas ratio"
- For an input interval [a, b] it is given by the area defined by intersection $\mu^0_{A_i} \cap \mu[a,b]$ divided by area defined by $\mu^0_{A_i}$

$$\mu_{A_i} = \frac{\int_a^b \min\left\{\mu^0_{A_i}(x), \mu_{[a,b]}(x)\right\} dx}{\int_a^b \mu^0_{A_i}(x) dx}$$

It is obvious that, any t-norm T can be used instead of min and its dual t-conorm S instead of *max* in the previous formulas.

Fuzzy Inference

The fuzzy control rules are of the form:

$$\mathrm{R_i : IF\ X_1\ is_i^{\,1}\ AND..AND\ X_r\ is\ A^r{}_i\ THEN\ Y\ is C_i}$$

where the variables $X_j\ j \in \{1,2,\ldots,r\}$, and Y have the domains u_j and V, respectively. The firing levels of the rules, denoted by $\{\alpha_i\}$, are computed by:

$$\alpha_i = T\left(\alpha_i^1,\ldots,\alpha_i^r\right)$$

where T is a t-norm and α_j^i is the firing level for $A_i^j, j \in \{1,2,\ldots,\}$. The causal link from $X_1,\ldots,X_r$ to Y is represented using an implication operator I. It results that the conclusion $C_i^{'}$ inferred from the rule R_i is:

$$\mu_{C_i^{'}}(v) = I\left(\alpha_i, \mu_{C_i}(v)\right), \forall v \in V.$$

The formula,

$$\mu_{C'}(v) = I(\alpha, \mu_C(v))$$

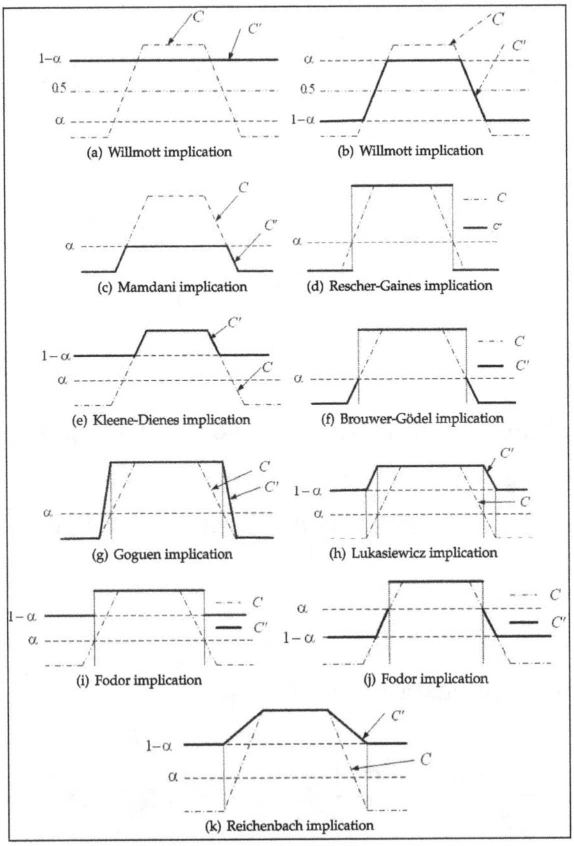

Conclusions obtained with different implications.

gives the following results, depending on the implication I:

$$\text{Willmott}: \mu_{C'}(v) = I_W(\alpha, \mu_C(v)) = \max\{1-\alpha, \min(\alpha \mu_C(v))\}$$

$$\text{Mamdani}: \mu_{C'}(v) = I_M(\alpha, \mu_C(v)) = \min\{\alpha, \mu_C(v)\}$$

$$\text{Rescher-Gaines}: \mu_{C'}(v) = I_{RG}(\alpha, \mu_C(v)) = \begin{cases} 1 & \text{if } \alpha \le \mu_C(v) \\ 0 & \text{otherwise} \end{cases}$$

$$\text{Kleene-Dienes}: \mu_{C'}(v) = I_{KD}(\alpha, \mu_C(v)) = \max\{1-\alpha, \mu_C(v)\}$$

$$\text{Brouwer-Gödel}: \mu_{C'}(v) = I_{BG}(\alpha, \mu_C(v)) = \begin{cases} 1 & \text{if } \alpha \le \mu_C(v) \\ \mu_C(v) & \text{otherwise} \end{cases}$$

$$\text{Goguen}: \mu_{C'}(v) = I_G(\alpha, \mu_C(v)) = \begin{cases} 1 \text{ if } \alpha \leq \mu_C(v) \\ \dfrac{\mu_C(v)}{\alpha} \text{ otherwise} \end{cases}$$

$$\text{Lukasiewicz}: \mu_{C'}(v) = I_L(\alpha, \mu_C(v)) = \min\{1 - \alpha + \mu_C(v), 1\}$$

$$\text{Fodor}: \mu_{C'}(v) = I_F(\alpha, \mu_C(v)) = \begin{cases} 1 \text{ if } \alpha \leq \mu_C(v) \\ \max\{1 - \alpha, \mu_C(v)\} \text{ otherwise} \end{cases}$$

$$\text{Reichenbach}: \mu_{C'}(v) = I_R(\alpha, \mu_C(v)) = 1 - \alpha + \alpha\mu_C(v)$$

Defuzzification

The fuzzy output C_i' of the rule R_i is transformed into a crisp output z_i using the Middle-of-Maxima operator. The crisp value z_0 associated to a conclusion C′ inferred from a rule having the firing level α and the conclusion C represented by the fuzzy number $(\underline{m}_C, \overline{m}_C, \alpha_C, \beta_C)$ is:

$$z_0 = \frac{\underline{m}_C + \overline{m}_C}{2} \text{ for implication } I \in \{I_R, I_{KD}\}$$

$$z_0 = \frac{\underline{m}_C + \overline{m}_C + (1-\alpha)(\beta_C - \alpha_C)}{2} \text{ for}$$

$$I \in \{I_M, I_{RG}, I_{BG}, I_G, I_L, I_F\} \text{ or } (I = I_W,\ \alpha \geq 0.5)$$

$$z_0 = \frac{a_V + b_V}{2} \text{ if } I = I_W,\ \alpha < 0.5 \text{ and } V = [a_V,\ b_V]$$

In the last case, in order to remain inside the support of C, one can choose a value according to Max-Criterion; for instance:

$$z_0 = \frac{\underline{m}_C + \overline{m}_C + \alpha(\beta_C - \alpha_C)}{2}$$

The overall crisp control action is computed by the discrete Center-of-Gravity method as follows. If the number of fired rules is N then the final control action is:

$$z_0 = \left(\sum_{i=1}^{N} \alpha_i z_i\right) / \sum_{i=1}^{N} \alpha_i$$

where α_i is the firing level and z_i is the crisp output of the i-th rule, $i = \overline{1, N}$.

Finally, the results obtained with various implication operators are combined in order to obtain the overall output of the system. For this reason, the "strength" λ(I) of an implication I is used:

$$\lambda(I) = N(I)/13$$

where N(I) is the number of properties (from the list I_1 to I_{13}) verified by the implication I.

$$w_1 = \lambda(I_W), w_2 = \lambda(I_M), \ldots, w_9 = \lambda(I_R)$$
$$a_1 = z_0(I_W),\ a_2 = z_0(I_M),\ \ldots,\ a_9 = z_0(I_R)$$

and the overall crisp action of the system is computed as:

$$\bar{z}_0 = \sum_{j=1}^{9} w_j b_j$$

where b_j is the j-th largest element of $\{z_0(I_W),\ z_0(I_M),\ \ldots,\ z_0(I_F),\ z_0(I_R)\}$.

References

- Fuzzy-number: whatis.techtarget.com, Retrieved 15 April, 2019
- Fuzzy-logic-set-theory, fuzzy-logic: tutorialspoint.com, Retrieved 14 February, 2019
- Mendel, Jerry M. (2001). Uncertain rule-based fuzzy logic systems : introduction and new directions. Upper Saddle River, NJ: Prentice Hall PTR. ISBN 978-0130409690. OCLC 45314121
- Michael Hanss, 2005. Applied Fuzzy Arithmetic, An Introduction with Engineering Applications. Springer, ISBN 3-540-24201-5
- Mamdani-fuzzy-model, engineering-fuzzy-system: researchhubs.com, Retrieved 27 July, 2019
- A-mamdani-type-fuzzy-logic-controller, fuzzy-logic-controls-concepts-theories-and-applications: intechopen.com, Retrieved 17 August, 2019

Formal Systems of Fuzzy Logic

CHAPTER 3

The formal system of fuzzy logic is categorized into propositional fuzzy logic and predicate fuzzy logic. It consists of T-norm fuzzy logic, Monoidal T-norm basic logic, basic fuzzy logic, Godel–Dummett logic and Lukasiewicz logic. The topics elaborated in this chapter will help in gaining a better perspective about these formal systems of fuzzy logic.

PROPOSITIONAL FUZZY LOGIC

The fuzzy propositional logic (FPL) presents a new alternative approach to thinking about such notions as a proposition, interpretation, inference. This logic generalizes the classical propositional logic in two directions: (i) propositions are considered to be fuzzy and (ii) logical variables are considered to be many-valued. Formulas of FPL are constructed from elementary propositions and interpreted over the universe, i.e. the Cartesian product of the sets of the values of all variables. The main inference rule in FPL is a fuzzy resolution which generalizes properties of the traditional resolution.

Propositional logics are concerned with propositional (or sentential) operators which may be applied to one or more propositions giving new propositions.

Accent on the truth value of propositions, not on their meaning, and on how these truth values are composed.

A logic is truth functional if the truth value of a compound sentence depends only on the truth values of the constituent atomic sentences, not on their meaning or structure. For such a logic the only important question about propositions is what truth values they may have.

In a classical, boolean or two-valued logic every proposition is either true or false and no other feature of the proposition is relevant.

Example of propositional logic. The conjunction of the two sentences:

- Grass is green.
- Pigs don't fly.

It is the sentence:

- Grass is green and pigs don't fly.

The conjunction of two sentences will be true if, and only if, each of the two sentences from which it was formed is true.

Other operators are disjunction (OR) and negation (NOT).

T-norm Fuzzy Logics

T-norm fuzzy logics are a family of non-classical logics, informally delimited by having a semantics that takes the real unit interval [0, 1] for the system of truth values and functions called t-norms for permissible interpretations of conjunction. They are mainly used in applied fuzzy logic and fuzzy set theory as a theoretical basis for approximate reasoning.

T-norm fuzzy logics belong in broader classes of fuzzy logics and many-valued logics. In order to generate a well-behaved implication, the t-norms are usually required to be left-continuous; logics of left-continuous t-norms further belong in the class of substructural logics, among which they are marked with the validity of the law of prelinearity, $(A \to B) \vee (B \to A)$. Both propositional and first-order (or higher-order) t-norm fuzzy logics, as well as their expansions by modal and other operators, are studied. Logics that restrict the t-norm semantics to a subset of the real unit interval (for example, finitely valued Łukasiewicz logics) are usually included in the class as well.

Important examples of t-norm fuzzy logics are monoidal t-norm logic MTL of all left-continuous t-norms, basic logic BL of all continuous t-norms, product fuzzy logic of the product t-norm, or the nilpotent minimum logic of the nilpotent minimum t-norm. Some independently motivated logics belong among t-norm fuzzy logics, too, for example Łukasiewicz logic (which is the logic of the Łukasiewicz t-norm) or Gödel–Dummett logic (which is the logic of the minimum t-norm).

As members of the family of fuzzy logics, t-norm fuzzy logics primarily aim at generalizing classical two-valued logic by admitting intermediary truth values between 1 (truth) and 0 (falsity) representing *degrees* of truth of propositions. The degrees are assumed to be real numbers from the unit interval [0, 1]. In propositional t-norm fuzzy logics, propositional connectives are stipulated to be truth-functional, that is, the truth value of a complex proposition formed by a propositional connective from some constituent propositions is a function (called the *truth function* of the connective) of the truth values of the constituent propositions. The truth functions operate on the set of truth degrees (in the standard semantics, on the [0, 1] interval); thus the truth function of an n-ary propositional connective c is a function $F_c\colon [0, 1]^n \to [0, 1]$. Truth functions generalize truth tables of propositional connectives known from classical logic to operate on the larger system of truth values.

T-norm fuzzy logics impose certain natural constraints on the truth function of conjunction. The truth function $*:[0,1]^2 \rightarrow [0,1]$ of conjunction is assumed to satisfy the following conditions:

- Commutativity, that is, $x * y = y * x$ for all x and y in [0, 1]. This expresses the assumption that the order of fuzzy propositions is immaterial in conjunction, even if intermediary truth degrees are admitted.
- Associativity, that is, $(x * y) * z = x * (y * z)$ for all x, y, and z in [0, 1]. This expresses the assumption that the order of performing conjunction is immaterial, even if intermediary truth degrees are admitted.
- Monotony, that is, if $x \leq y$ then $x * z \leq y * z$ for all x, y, and z in [0, 1]. This expresses the assumption that increasing the truth degree of a conjunct should not decrease the truth degree of the conjunction.
- Neutrality of 1, that is, $1 * x = x$ for all x in [0, 1]. This assumption corresponds to regarding the truth degree 1 as full truth, conjunction with which does not decrease the truth value of the other conjunct. Together with the previous conditions this condition ensures that also $0 * x = 0$ for all x in [0, 1], which corresponds to regarding the truth degree 0 as full falsity, conjunction with which is always fully false.
- Continuity of the function $*$ (the previous conditions reduce this requirement to the continuity in either argument). Informally this expresses the assumption that microscopic changes of the truth degrees of conjuncts should not result in a macroscopic change of the truth degree of their conjunction. This condition, among other things, ensures a good behavior of (residual) implication derived from conjunction; to ensure the good behavior, however, *left*-continuity (in either argument) of the function $*$ is sufficient. In general t-norm fuzzy logics, therefore, only left-continuity of $*$ is required, which expresses the assumption that a microscopic *decrease* of the truth degree of a conjunct should not macroscopically decrease the truth degree of conjunction.

These assumptions make the truth function of conjunction a left-continuous t-norm, which explains the name of the family of fuzzy logics (*t-norm based*). Particular logics of the family can make further assumptions about the behavior of conjunction (for example, Gödel logic requires its idempotence) or other connectives (for example, the logic IMTL requires the involutiveness of negation).

All left-continuous t-norms $*$ have a unique residuum, that is, a binary function $\Rightarrow$ such that for all x, y, and z in [0, 1],

$x * y \leq z$ if and only if $x \leq y \Rightarrow z$.

The residuum of a left-continuous t-norm can explicitly be defined as:

$$(x \Rightarrow y) = \sup\{z \mid z * x \leq y\}$$

This ensures that the residuum is the pointwise largest function such that for all x and y,

$$x * (x \Rightarrow y) \leq y.$$

The latter can be interpreted as a fuzzy version of the modus ponens rule of inference. The residuum of a left-continuous t-norm thus can be characterized as the weakest function that makes the fuzzy modus ponens valid, which makes it a suitable truth function for implication in fuzzy logic. Left-continuity of the t-norm is the necessary and sufficient condition for this relationship between a t-norm conjunction and its residual implication to hold.

Truth functions of further propositional connectives can be defined by means of the t-norm and its residuum, for instance the residual negation $\neg x = (x \Rightarrow 0)$ or bi-residual equivalence $x \Leftrightarrow y = (x \Rightarrow y) * (y \Rightarrow x)$ Truth functions of propositional connectives may also be introduced by additional definitions: the most usual ones are the minimum (which plays a role of another conjunctive connective), the maximum (which plays a role of a disjunctive connective), or the Baaz Delta operator, defined in [0, 1] as $\Delta x = 1$ if $x = 1$ and $\Delta x = 0$ otherwise. In this way, a left-continuous t-norm, its residuum, and the truth functions of additional propositional connectives determine the truth values of complex propositional formulae in [0, 1].

Formulae that always evaluate to 1 are called *tautologies* with respect to the given left-continuous t-norm $*$, or $*$–*tautologies*. The set of all $*$–tautologies is called the *logic* of the t-norm $*$, as these formulae represent the laws of fuzzy logic (determined by the t-norm) which hold (to degree 1) regardless of the truth degrees of atomic formulae. Some formulae are tautologies with respect to a larger class of left-continuous t-norms; the set of such formulae is called the logic of the class. Important t-norm logics are the logics of particular t-norms or classes of t-norms, for example:

- Łukasiewicz logic is the logic of the Łukasiewicz t-norm $x * y = \max(x + y - 1, 0)$.
- Gödel–Dummett logic is the logic of the minimum t-norm $x * y = \min(x, y)$.
- Product fuzzy logic is the logic of the product t-norm $x * y = x \cdot y$.
- Monoidal t-norm logic MTL is the logic of (the class of) *all* left-continuous t-norms.
- Basic fuzzy logic BL is the logic of (the class of) all *continuous* t-norms.

It turns out that many logics of particular t-norms and classes of t-norms are axiomatizable. The completeness theorem of the axiomatic system with respect to the corresponding t-norm semantics on [0, 1] is then called the *standard completeness* of the logic. Besides the standard real-valued semantics on [0, 1], the logics are sound and complete with respect to general algebraic semantics, formed by suitable classes of prelinear commutative bounded integral residuated lattices.

Some particular t-norm fuzzy logics have been introduced and investigated long before the family was recognized (even before the notions of fuzzy logic or t-norm emerged):

- Łukasiewicz logic (the logic of the Łukasiewicz t-norm) was originally defined by Jan Łukasiewicz as a three-valued logic; it was later generalized to n-valued (for all finite n) as well as infinitely-many-valued variants, both propositional and first-order.

- Gödel–Dummett logic (the logic of the minimum t-norm) was implicit in Gödel's proof of infinite-valuedness of intuitionistic logic. Later it was explicitly studied by Dummett who proved a completeness theorem for the logic.

A systematic study of particular t-norm fuzzy logics and their classes began with Hájek's monograph Metamathematics of Fuzzy Logic, which presented the notion of the logic of a continuous t-norm, the logics of the three basic continuous t-norms (Łukasiewicz, Gödel, and product), and the 'basic' fuzzy logic BL of all continuous t-norms (all of them both propositional and first-order). The book also started the investigation of fuzzy logics as non-classical logics with Hilbert-style calculi, algebraic semantics, and metamathematical properties known from other logics (completeness theorems, deduction theorems, complexity, etc.).

Since then, a plethora of t-norm fuzzy logics have been introduced and their metamathematical properties have been investigated. Some of the most important t-norm fuzzy logics were introduced in 2001, by Esteva and Godo (MTL, IMTL, SMTL, NM, WNM), Esteva, Godo, and Montagna (propositional ŁΠ), and Cintula (first-order ŁΠ).

Logical Language

The logical vocabulary of propositional t-norm fuzzy logics standardly comprises the following connectives:

- Implication $\rightarrow$ (binary). In the context of other than t-norm-based fuzzy logics, the t-norm-based implication is sometimes called residual implication or R-implication, as its standard semantics is the residuum of the t-norm that realizes strong conjunction.

- Strong conjunction & (binary). In the context of substructural logics, the sign $\otimes$ and the names *group, intensional, multiplicative,* or *parallel conjunction* are often used for strong conjunction.

- Weak conjunction $\wedge$ (binary), also called lattice conjunction (as it is always realized by the lattice operation of meet in algebraic semantics). In the context of substructural logics, the names *additive, extensional,* or *comparative conjunction* are sometimes used for lattice conjunction. In the logic BL and its extensions (though not in t-norm logics in general), weak conjunction is definable in terms of implication and strong conjunction, by,

$$A \wedge B \equiv A \,\&\, (A \rightarrow B)$$

The presence of two conjunction connectives is a common feature of contraction-free substructural logics.

- Bottom $\perp$ (nullary); 0 or $\overline{0}$ are common alternative signs and zero a common alternative name for the propositional constant (as the constants bottom and zero of substructural logics coincide in t-norm fuzzy logics). The proposition $\perp$ represents the *falsity* or *absurdum* and corresponds to the classical truth value *false*.
- Negation $\neg$ (unary), sometimes called residual negation if other negation connectives are considered, as it is defined from the residual implication by the reductio ad absurdum:

  $$\neg A \equiv A \rightarrow \perp$$

- Equivalence $\leftrightarrow$ (binary), defined as:

  $$A \leftrightarrow B \equiv (A \rightarrow B) \wedge (B \rightarrow A)$$

In t-norm logics, the definition is equivalent to $(A \rightarrow B) \,\&\, (B \rightarrow A)$.

- (Weak) disjunction $\vee$ (binary), also called lattice disjunction (as it is always realized by the lattice operation of join in algebraic semantics). In t-norm logics it is definable in terms of other connectives as:

  $$A \vee B \equiv ((A \rightarrow B) \rightarrow B) \wedge ((B \rightarrow A) \rightarrow A)$$

- Top $\top$ (nullary), also called one and denoted by 1 or $\overline{1}$ (as the constants top and zero of substructural logics coincide in t-norm fuzzy logics). The proposition $\top$ corresponds to the classical truth value *true* and can in t-norm logics be defined as:

  $$\top \equiv \perp \rightarrow \perp.$$

Some propositional t-norm logics add further propositional connectives to the above language, most often the following ones:

- The Delta connective $\vartriangle$ is a unary connective that asserts classical truth of a proposition, as the formulae of the form $\vartriangle A$ behave as in classical logic. Also called the Baaz Delta, as it was first used by Matthias Baaz for Gödel–Dummett logic. The expansion of a t-norm logic L by the Delta connective is usually denoted by $L_{\vartriangle}$.

- Truth constants are nullary connectives representing particular truth values between 0 and 1 in the standard real-valued semantics. For the real number r, the corresponding truth constant is usually denoted by $\bar{r}$ Most often, the truth constants for all rational numbers are added. The system of all truth constants in the language is supposed to satisfy the *bookkeeping axioms*: $\overline{r \& s} \leftrightarrow (\bar{r} \& \bar{s}), \overline{r \rightarrow s} \leftrightarrow (\bar{r} \rightarrow \bar{s})$, etc. for all propositional connectives and all truth constants definable in the language.

- Involutive negation $\sim$ (unary) can be added as an additional negation to t-norm logics whose residual negation is not itself involutive, that is, if it does not obey the law of double negation $\neg\neg A \leftrightarrow A$. A t-norm logic L expanded with involutive negation is usually denoted by $L_{\sim}$ and called *L with involution.*

- Strong disjunction $\oplus$ (binary). In the context of substructural logics it is also called *group, intensional, multiplicative*, or *parallel disjunction*. Even though standard in contraction-free substructural logics, in t-norm fuzzy logics it is usually used only in the presence of involutive negation, which makes it definable (and so axiomatizable) by de Morgan's law from strong conjunction:

  $A \oplus B \equiv \sim (\sim A \& \sim B)$

- Additional t-norm conjunctions and residual implications. Some expressively strong t-norm logics, for instance the logic ŁΠ, have more than one strong conjunction or residual implication in their language. In the standard real-valued semantics, all such strong conjunctions are realized by different t-norms and the residual implications by their residua.

Well-formed formulae of propositional t-norm logics are defined from propositional variables (usually countably many) by the above logical connectives, as usual in propositional logics. In order to save parentheses, it is common to use the following order of precedence:

- Unary connectives (bind most closely).
- Binary connectives other than implication and equivalence.
- Implication and equivalence (bind most loosely).

First-order variants of t-norm logics employ the usual logical language of first-order logic with the above propositional connectives and the following quantifiers:

- General quantifier $\forall$.
- Existential quantifier $\exists$.

The first-order variant of a propositional t-norm logic L is usually denoted by $L\forall$.

Semantics

Algebraic semantics is predominantly used for propositional t-norm fuzzy logics, with three main classes of algebras with respect to which a t-norm fuzzy logic L is complete:

- General semantics, formed of all *L-algebras* — that is, all algebras for which the logic is sound.
- Linear semantics, formed of all *linear* L-algebras — that is, all L-algebras whose lattice order is linear.
- Standard semantics, formed of all *standard* L-algebras — that is, all L-algebras whose lattice reduct is the real unit interval [0, 1] with the usual order. In standard L-algebras, the interpretation of strong conjunction is a left-continuous t-norm and the interpretation of most propositional connectives is determined by the t-norm (hence the names *t-norm-based logics* and *t-norm* $L_{\sim}$*-algebras*, which is also used for L-algebras on the lattice [0, 1]). In t-norm logics with additional connectives, however, the real-valued interpretation of the additional connectives may be restricted by further conditions for the t-norm algebra to be called standard: for example, in standard $L_{\sim}$-algebras of the logic L with involution, the interpretation of the additional involutive negation $\sim$ is required to be the *standard involution* $f_{\sim}(x) = 1 - x$, rather than other involutions which can also interpret $\sim$ over t-norm $L_{\sim}$-algebras. In general, therefore, the definition of standard t-norm algebras has to be explicitly given for t-norm logics with additional connectives.

Monoidal T-norm Logic

Monoidal t-norm based logic (or shortly MTL), the logic of left-continuous t-norms, is one of t-norm fuzzy logics. It belongs to the broader class of substructural logics, or logics of residuated lattices; it extends the logic of commutative bounded integral residuated lattices (known as Höhle's monoidal logic, Ono's FL_{ew}, or intuitionistic logic without contraction) by the axiom of prelinearity.

In fuzzy logic, rather than regarding statements as being either true or false, we associate each statement with a numerical *confidence* in that statement. By convention the confidences range over the unit interval **[0,1]**, where the maximal confidence **1** corresponds to the classical concept of true and the minimal confidence 0 corresponds to the classical concept of false.

T-norms are binary functions on the real unit interval [0, 1] which in fuzzy logic are often used to represent a conjunction connective; if $a, b \in [0,1]$ are the confidences we ascribe to the statements **A** and **B** respectively, then one uses a t-norm $*$ to calculate the confidence $\boldsymbol{a} * \boldsymbol{b}$ ascribed to the compound statement '**A** and **B**'. A t-norm * has to satisfy the properties of:

- Commutativity $a * b = b * a$,
- Associativity $(a*b)*c = a*(b*c)$,
- Monotonicity – if $a \le b$ and $c \le d$ then $a*c \le b*d$,
- And having 1 as identity element $1*a = a$.

Notably absent from this list is the property of idempotence $\boldsymbol{a} * \boldsymbol{a} = \boldsymbol{a}$; the closest one gets is that $\boldsymbol{a} * \boldsymbol{a} \le \mathbf{1} * \boldsymbol{a} = \boldsymbol{a}$. It may seem strange to be less confident in '**A** and **A**' than in just **A**, but we generally do want to allow for letting the confidence $\boldsymbol{a} * \boldsymbol{b}$ in a combined '**A** and **B**' be less than both the confidence **a** in **A** and the confidence **b** in **B**, and then the ordering $\boldsymbol{a} * \boldsymbol{b} < \mathbf{a} \le \mathbf{b}$ by monotonicity requires $\boldsymbol{a} * \boldsymbol{a} \le \mathbf{a} * \mathbf{b} < \mathbf{a}$. Another way of putting it is that the t-norm can only take into account the confidences as numbers, not the reasons that may be behind ascribing those confidences; thus it cannot treat '**A** and **A**' differently from '**A** and **B**' where we are equally confident in both'.

Because the symbol $\wedge$ via its use in lattice theory is very closely associated with the idempotence property, it can be useful to switch to a different symbol for conjunction that is not necessarily idempotent. In the fuzzy logic tradition one sometimes uses & for this "strong" conjunction, but the substructural logic tradition of using $\otimes$ for the strong conjunction; thus $\boldsymbol{a} * \boldsymbol{b}$ is the confidence we ascribe to the statement $\boldsymbol{A} \otimes \boldsymbol{B}$ (still read '**A** and **B**', perhaps with 'strong' or 'multiplicative' as qualification of the 'and').

Having formalised conjunction $\otimes$, one wishes to continue with the other connectives. One approach to doing that is to introduce negation as an order-reversing map $[0,1] \to [0,1]$, then defining remaining connectives using De Morgan's laws, material implication, and the like. A problem with doing so is that the resulting logics may have undesirable properties: they may be too close to classical logic, or if not conversely not support expected inference rules. An alternative that makes the consequences of different choices more predictable is to instead continue with implication $\to$ as the second connective: this is overall the most common connective in axiomatisations of logic, and it has closer ties to the deductive aspects of logic than most other connectives. A confidence counterpart $\Rightarrow$ of the implication connective may in fact be defined directly as the residuum of the t-norm.

The logical link between conjunction and implication is provided by something as fundamental as the inference rule modus ponens $\boldsymbol{A}, \boldsymbol{A} \to \boldsymbol{B} \vdash \boldsymbol{B}$: from **A** and $\boldsymbol{A} \to \boldsymbol{B}$ follows **B**. In the fuzzy logic case that is more rigorously written as $\boldsymbol{A} \otimes (\boldsymbol{A} \to \boldsymbol{B}) \vdash \boldsymbol{B}$, because this makes explicit that our confidence for the premises here is that in $\boldsymbol{A} \otimes (\boldsymbol{A} \to \boldsymbol{B})$, not those in **A** and $\boldsymbol{A} \to \boldsymbol{B}$ separately. So if **a** and **b** are our confidences in

A and **B** respectively, then $a \Rightarrow b$ is the sought confidence in $A \to B$, and $a*(a \Rightarrow b)$ is the combined confidence in $A \otimes (A \to B)$. We require that:

$$a*(a \Rightarrow b) \leq b$$

since our confidence **b** for **B** should not be less than our confidence $a*(a \Rightarrow b)$ in the statement $A \otimes (A \to B)$ from which **B** logically follows. This bounds the sought confidence $a \Rightarrow b$, and one approach for turning $\Rightarrow$ into a binary operation like $*$ would be to make it as large as possible while respecting this bound:

$$a \Rightarrow b \equiv \sup\left\{x \in [0,1] \middle| a*x \leq b\right\}.$$

Taking $x = 0$ gives $a*x = a*0 \leq 1*0 = 0 \leq b$, so the supremum is always of a nonempty bounded set and thus well-defined. For a general t-norm there remains the possibility that $f_a(x) = a*x$ has a jump discontinuity at $x = a \Rightarrow b$, in which case $a*(a \Rightarrow b)$ could come out strictly larger than b even though $a \Rightarrow b$ is defined as the least upper bound of xs satisfying $a*x \leq b$; to prevent that and have the construction work as expected, we require that the t-norm * is left-continuous. The residuum of a left-continuous t-norm thus can be characterized as the weakest function that makes the fuzzy modus ponens valid, which makes it a suitable truth function for implication in fuzzy logic.

More algebraically, we say that an operation $\Rightarrow$ is a residuum of a t-norm $*$ if for all **a**, **b**, and **c** it satisfies:

$$a*b \leq c \text{ if and only if } a \leq (b \Rightarrow c).$$

This equivalence of numerical comparisons mirrors the equivalence of entailments:

$$A \otimes B \vdash C \text{ if and only if } A \vdash B \to C$$

That exists because any proof of **C** from the premise $A \otimes B$ can be converted into a proof of $B \to C$ from the premise **A** by doing an extra implication introduction step, and conversely any proof of $B \to C$ from the premise **A** can be converted into a proof of **C** from the premise $A \otimes B$ by doing an extra implication elimination step. Left-continuity of the t-norm is the necessary and sufficient condition for this relationship between a t-norm conjunction and its residual implication to hold.

Truth functions of further propositional connectives can be defined by means of the t-norm and its residuum, for instance the residual negation $\neg x = (x \Rightarrow 0)$ In this way, the left-continuous t-norm, its residuum, and the truth functions of additional propositional connectives determine the truth values of complex propositional formulae in [0, 1]. Formulae that always evaluate to 1 are then called *tautologies* with respect to the given left-continuous t-norm *, or **-tautologies*. The set of all *- tautologies is called the *logic* of the t-norm *, since these formulae represent the laws of fuzzy logic (determined by the t-norm) which hold (to degree 1) regardless of the truth degrees of atomic formulae. Some formulae are tautologies with respect to *all* left-continuous t-norms: they represent general laws of propositional fuzzy logic which are independent of the choice of a particular left-continuous t-norm. These formulae form the logic MTL, which can thus be characterized as the *logic of left-continuous t-norms.*

Syntax

Language

The language of the propositional logic MTL consists of countably many propositional variables and the following primitive logical connectives:

- Implication $\rightarrow$ (binary).
- Strong conjunction $\otimes$ (binary). The sign & is a more traditional notation for strong conjunction in the literature on fuzzy logic, while the notation $\otimes$ follows the tradition of substructural logics.
- Weak conjunction $\wedge$ (binary), also called lattice conjunction (as it is always realized by the lattice operation of meet in algebraic semantics). Unlike BL and stronger fuzzy logics, weak conjunction is not definable in MTL and has to be included among primitive connectives.
- Bottom $\bot$ (nullary — a propositional constant; 0 or $\bar{0}$ are common alternative tokens and zero a common alternative name for the propositional constant (as the constants *bottom* and *zero* of substructural logics coincide in MTL).
- The following are the most common defined logical connectives:
- Negation $\neg$ (unary), defined as:

  $$\neg A \equiv A \rightarrow \bot$$

- Equivalence $\leftrightarrow$ (binary), defined as:

  $$A \leftrightarrow B \equiv (A \rightarrow B) \wedge (B \rightarrow A)$$

In MTL, the definition is equivalent to $(A \rightarrow B) \otimes (B \rightarrow A)$.

- (Weak) disjunction $\vee$ (binary), also called lattice disjunction (as it is always realized by the lattice operation of join in algebraic semantics), defined as:

  $$A \vee B \equiv ((A \rightarrow B) \rightarrow B) \wedge ((B \rightarrow A) \rightarrow A)$$

- Top $\top$ (nullary), also called one and denoted by 1 or $\bar{1}$ (as the constants top and zero of substructural logics coincide in MTL), defined as:

  $$\top \equiv \bot \rightarrow \bot$$

Well-formed formulae of MTL are defined as usual in propositional logics. In order to save parentheses, it is common to use the following order of precedence:

- Unary connectives (bind most closely)
- Binary connectives other than implication and equivalence
- Implication and equivalence (bind most loosely)

Axioms

A Hilbert-style deduction system for MTL has been introduced by Esteva and Godo (2001). Its single derivation rule is modus ponens:

from A and $A \rightarrow B$ derive B.

The following are its axiom schemata:

$$
\begin{array}{ll}
(\mathrm{MTL1}): & (A \rightarrow B) \rightarrow ((B \rightarrow C) \rightarrow (A \rightarrow C)) \\
(\mathrm{MTL2}): & A \otimes B \rightarrow A \\
(\mathrm{MTL3}): & A \otimes B \rightarrow B \otimes A \\
(\mathrm{MTL4a}): & A \otimes B \rightarrow B \otimes A \\
(\mathrm{MTL4b}): & A \wedge B \rightarrow B \wedge A \\
(\mathrm{MTL4c}): & A \otimes (A \rightarrow B) \rightarrow A \wedge B \\
(\mathrm{MTL5a}): & (A \rightarrow (B \rightarrow C)) \rightarrow (A \otimes B \rightarrow C) \\
(\mathrm{MTL5b}): & (A \otimes B \rightarrow C) \rightarrow (A \rightarrow (B \rightarrow C)) \\
(\mathrm{MTL6}): & ((A \rightarrow B) \rightarrow C) \rightarrow (((B \rightarrow A) \rightarrow C) \rightarrow C) \\
(\mathrm{MTL7}): & \bot \rightarrow A
\end{array}
$$

The traditional numbering of axioms, given in the left column, is derived from the numbering of axioms of Hájek's basic fuzzy logic BL. The axioms (MTL4a)–(MTL4c) replace the axiom of *divisibility* (BL4) of BL. The axioms (MTL5a) and (MTL5b) express the law of residuation and the axiom (MTL6) corresponds to the condition of prelinearity. The axioms (MTL2) and (MTL3) of the original axiomatic system were shown to be redundant and. All the other axioms were shown to be independent.

Semantics

Like in other propositional t-norm fuzzy logics, algebraic semantics is predominantly used for MTL, with three main classes of algebras with respect to which the logic is complete:

- General semantics, formed of all *MTL-algebras* — that is, all algebras for which the logic is sound.

- Linear semantics, formed of all *linear* MTL-algebras — that is, all MTL-algebras whose lattice order is linear.
- Standard semantics, formed of all *standard* MTL-algebras — that is, all MTL-algebras whose lattice reduct is the real unit interval [0, 1] with the usual order; they are uniquely determined by the function that interprets strong conjunction, which can be any left-continuous t-norm.

General Semantics

MTL-algebras

Algebras for which the logic MTL is sound are called *MTL-algebras*. They can be characterized as *prelinear commutative bounded integral residuated lattices*. In more detail, an algebraic structure $(L, \wedge, \vee, *, \Rightarrow, 0, 1)$ is an MTL-algebra if:

- $(L, \wedge, \vee, 0, 1)$ is a bounded lattice with the top element 0 and bottom element 1.
- $(L, *, 1)$ is a commutative monoid.
- $*$ and $\Rightarrow$ form an adjoint pair, that is, $z * x \leq y$ if and only if $z \leq x \Rightarrow y$, where $\leq$ is the lattice order of $(L, \wedge, \vee)$, for all *x*, *y*, and *z* in L, (the *residuation* condition).
- $(x \Rightarrow y) \vee (y \Rightarrow x) = 1$ holds for all *x* and *y* in *L* (the *prelinearity* condition).

Important examples of MTL algebras are *standard* MTL-algebras on the real unit interval [0, 1]. Further examples include all Boolean algebras, all linear Heyting algebras (both with $* = \wedge$), all MV-algebras, all BL-algebras, etc. Since the residuation condition can equivalently be expressed by identities, MTL-algebras form a variety.

Interpretation of the Logic MTL in MTL-algebras

The connectives of MTL are interpreted in MTL-algebras as follows:

- Strong conjunction by the monoidal operation $*$.
- Implication by the operation $\Rightarrow$ (which is called the *residuum* of $*$).
- Weak conjunction and weak disjunction by the lattice operations $\wedge$ and $\vee$ respectively (usually denoted by the same symbols as the connectives, if no confusion can arise)
- The truth constants zero (top) and one (bottom) by the constants 0 and 1.
- The equivalence connective is interpreted by the operation $\Leftrightarrow$ defined as:

  $x \Leftrightarrow y \equiv (x \Rightarrow y) \wedge (y \Rightarrow x)$

Due to the prelinearity condition, this definition is equivalent to one that uses * instead of $\wedge$ thus,

$$x \Leftrightarrow y \equiv (x \Rightarrow y) * (y \Rightarrow x)$$

- Negation is interpreted by the definable operation $-x \equiv x \Rightarrow 0$.

With this interpretation of connectives, any evaluation e_v of propositional variables in L uniquely extends to an evaluation e of all well-formed formulae of MTL, by the following inductive definition (which generalizes Tarski's truth conditions), for any formulae A, B, and any propositional variable p:

$$\begin{array}{rcl} e(p) & = & e_v(p) \\ e(\bot) & = & 0 \\ e(\top) & = & 1 \\ e(A \otimes B) & = & e(A) * e(B) \\ e(A \to B) & = & e(A) \Rightarrow e(B) \\ e(A \wedge B) & = & e(A) \wedge e(B) \\ e(A \vee B) & = & e(A) \vee e(B) \\ e(A \leftrightarrow B) & = & e(A) \Leftrightarrow e(B) \\ e(\neg A) & = & e(A) \Rightarrow 0 \end{array}$$

Informally, the truth value 1 represents full truth and the truth value 0 represents full falsity; intermediate truth values represent intermediate degrees of truth. Thus a formula is considered fully true under an evaluation e if $e(A) = 1$. A formula A is said to be *valid* in an MTL-algebra L if it is fully true under all evaluations in L, that is, if $e(A) = 1$ for all evaluations e in L. Some formulae (for instance, $p \to p$) are valid in any MTL-algebra; these are called *tautologies* of MTL.

The notion of global entailment (or: global consequence) is defined for MTL as follows: a set of formulae Γ entails a formula A (or: A is a global consequence of Γ), in symbols $\Gamma \models A$, if for any evaluation e in any MTL-algebra, whenever $e(B) = 1$ for all formulae B in Γ, then also $e(A) = 1$. Informally, the global consequence relation represents the transmission of full truth in any MTL-algebra of truth values.

General Soundness and Completeness Theorems

The logic MTL is sound and complete with respect to the class of all MTL-algebras:

> A formula is provable in MTL if and only if it is valid in all MTL-algebras.

The notion of MTL-algebra is in fact so defined that MTL-algebras form the class of *all* algebras for which the logic MTL is sound. Furthermore, the *strong completeness theorem* holds:

> A formula A is a global consequence in MTL of a set of formulae Γ if and only if A is derivable from Γ in MTL.

Linear Semantics

Like algebras for other fuzzy logics, MTL-algebras enjoy the following *linear subdirect decomposition property*:

Every MTL-algebra is a subdirect product of linearly ordered MTL-algebras.

(A subdirect product is a subalgebra of the direct product such that all projection maps are surjective. An MTL-algebra is linearly ordered if its lattice order is linear.)

In consequence of the linear subdirect decomposition property of all MTL-algebras, the completeness theorem with respect to linear MTL-algebras holds:

- A formula is provable in MTL if and only if it is valid in all linear MTL-algebras.
- A formula A is derivable in MTL from a set of formulae Γ if and only if A is a global consequence in all linear MTL-algebras of Γ.

Standard Semantics

Standard are called those MTL-algebras whose lattice reduct is the real unit interval [0, 1]. They are uniquely determined by the real-valued function that interprets strong conjunction, which can be any left-continuous t-norm $*$. The standard MTL-algebra determined by a left-continuous t-norm $*$ is usually denoted by $[0,1]_*$. In $[0,1]_*$, implication is represented by the residuum of $*$, weak conjunction and disjunction respectively by the minimum and maximum, and the truth constants zero and one respectively by the real numbers 0 and 1.

The logic MTL is complete with respect to standard MTL-algebras; this fact is expressed by the standard completeness theorem:

A formula is provable in MTL if and only if it is valid in all standard MTL-algebras.

Since MTL is complete with respect to standard MTL-algebras, which are determined by left-continuous t-norms, MTL is often referred to as the logic of left-continuous t-norms (similarly as BL is the logic of continuous t-norms).

Basic Fuzzy Logic

Basic fuzzy Logic (or shortly BL), the logic of continuous t-norms, is one of t-norm fuzzy logics. It belongs to the broader class of substructural logics, or logics of residuated lattices; it extends the logic of all left-continuous t-norms MTL.

Syntax

Language

The language of the propositional logic BL consists of countably many propositional variables and the following primitive logical connectives:

- Implication $\rightarrow$ (binary).

- Strong conjunction $\otimes$ (binary). The sign & is a more traditional notation for strong conjunction in the literature on fuzzy logic, while the notation $\otimes$ follows the tradition of substructural logics.
- Bottom $\bot$ (nullary – a propositional constant); 0 or $\bar{0}$ are common alternative signs and zero a common alternative name for the propositional constant (as the constants bottom and zero of substructural logics coincide in MTL).

The following are the most common defined logical connectives:

- Weak conjunction $\wedge$ (binary), also called lattice conjunction (as it is always realized by the lattice operation of meet in algebraic semantics). Unlike MTL and weaker substructural logics, weak conjunction is definable in BL as:

  $$A \wedge B \equiv A \otimes (A \to B)$$

- Negation $\neg$ (unary), defined as:

  $$\neg A \equiv A \to \bot$$

- Equivalence $\leftrightarrow$ (binary), defined as:

  $$A \leftrightarrow B \equiv (A \to B) \wedge (B \to A)$$

As in MTL, the definition is equivalent to $(A \to B) \otimes (B \to A)$.

- (Weak) disjunction $\vee$ (binary), also called lattice disjunction (as it is always realized by the lattice operation of join in algebraic semantics), defined as:

  $$A \vee B \equiv ((A \to B) \to B) \wedge ((B \to A) \to A)$$

- Top $\top$ (nullary), also called one and denoted by 1 or $\bar{1}$ (as the constants top and zero of substructural logics coincide in MTL), defined as:

  $$\top \equiv \bot \to \bot$$

Well-formed formulae of BL are defined as usual in propositional logics. In order to save parentheses, it is common to use the following order of precedence:

- Unary connectives (bind most closely).
- Binary connectives other than implication and equivalence.
- Implication and equivalence (bind most loosely).

Axioms

A Hilbert-style deduction system for BL has been introduced by Petr Hájek. Its single derivation rule is modus ponens:

from **A** and $A \to B$ derive **B.**

The following are its axiom schemata:

$$
\begin{array}{ll}
(\text{BL1}): & (A \to B) \to ((B \to C) \to (A \to C)) \\
(\text{BL2}): & A \otimes B \to A \\
(\text{BL3}): & A \otimes B \to B \otimes A \\
(\text{BL4}): & A \otimes (A \to B) \to B \otimes (B \to A) \\
(\text{BL5a}): & (A \to (B \to C)) \to (A \otimes B \to C) \\
(\text{BL5b}): & (A \otimes B \to C) \to (A \to (B \to C)) \\
(\text{BL6}): & ((A \to B) \to C) \to (((B \to A) \to C) \to C) \\
(\text{BL7}): & \bot \to A
\end{array}
$$

The axioms (BL2) and (BL3) of the original axiomatic system were shown to be redundant. All the other axioms were shown to be independent.

Semantics

Like in other propositional t-norm fuzzy logics, algebraic semantics is predominantly used for BL, with three main classes of algebras with respect to which the logic is complete:

- General semantics, formed of all *BL-algebras* — that is, all algebras for which the logic is sound.
- Linear semantics, formed of all *linear* BL-algebras — that is, all BL-algebras whose lattice order is linear.
- Standard semantics, formed of all *standard* BL-algebras — that is, all BL-algebras whose lattice reduct is the real unit interval [0, 1] with the usual order; they are uniquely determined by the function that interprets strong conjunction, which can be any continuous t-norm.

Łukasiewicz Logic

Łukasiewicz logic is a non-classical, many-valued logic. It was originally defined in the early 20th century by Jan Łukasiewicz as a three-valued logic; it was later generalized to n-valued (for all finite n) as well as infinitely-many-valued ($\aleph_0$-valued) variants, both propositional and first-order. The $\aleph_0$-valued version was published in 1930 by Łukasiewicz and Alfred Tarski; consequently it is sometimes called the Łukasiewicz–

Tarski logic. It belongs to the classes of t-norm fuzzy logics and substructural logics.

Language

The propositional connectives of Łukasiewicz logic are *implication* $\rightarrow$, *negation* $\neg$, *equivalence* $\leftrightarrow$, *weak conjunction* $\wedge$, *strong conjunction* $\otimes$, *weak disjunction* $\vee$, *strong disjunction* $\otimes$, and propositional constants $\overline{0}$ and $\overline{1}$. The presence of conjunction and disjunction is a common feature of substructural logics without the rule of contraction, to which Łukasiewicz logic belongs.

The original system of axioms for propositional infinite-valued Łukasiewicz logic used implication and negation as the primitive connectives:

$$A \rightarrow (B \rightarrow A)$$
$$(A \rightarrow B) \rightarrow ((B \rightarrow C) \rightarrow (A \rightarrow C))$$
$$((A \rightarrow B) \rightarrow B) \rightarrow ((B \rightarrow A) \rightarrow A)$$
$$(\neg B \rightarrow \neg A) \rightarrow (A \rightarrow B)$$

Propositional infinite-valued Łukasiewicz logic can also be axiomatized by adding the following axioms to the axiomatic system of monoidal t-norm logic:

Divisibility

$$(A \wedge B) \rightarrow (A \otimes (A \rightarrow B))$$

Double Negation

$$\neg\neg A \rightarrow A.$$

That is, infinite-valued Łukasiewicz logic arises by adding the axiom of double negation to basic t-norm logic BL, or by adding the axiom of divisibility to the logic IMTL.

Finite-valued Łukasiewicz logics require additional axioms.

Real-valued Semantics

Infinite-valued Łukasiewicz logic is a real-valued logic in which sentences from sentential calculus may be assigned a truth value of not only zero or one but also any real number in between (e.g. 0.25). Valuations have a recursive definition where:

- $w(\theta \circ \phi) = F_{\circ}(w(\theta), w(\phi))$ for a binary connective o,
- $w(\neg\theta) = F_{\neg}(w(\theta))$,
- $w(\overline{0}) = 0$ and $w(\overline{1}) = 1$

And where the definitions of the operations hold as follows:

- Implication: $F_{\rightarrow}(x, y) = \min\{1, 1 - x + y\}$
- Equivalence: $F_{\leftrightarrow}(x, y) = 1 - |x - y|$
- Negation: $F_{\neg}(x) = 1 - x$
- Weak conjunction: $F_{\wedge}(x, y) \quad \min\{x, y\}$
- Weak disjunction: $F_{\vee}(x, y) = \max\{x, y\}$
- Strong conjunction: $F_{\otimes}(x, y) = \max\{0, x + y - 1\}$
- Strong disjunction: $F_{\oplus}(x, y) = \min\{1, x + y\}$

The truth function $F_{\otimes}$ of strong conjunction is the Łukasiewicz t-norm and the truth function $F_{\otimes}$ of strong disjunction is its dual t-conorm. The truth function $F_{\rightarrow}$ is the residuum of the Łukasiewicz t-norm. All truth functions of the basic connectives are continuous.

By definition, a formula is a tautology of infinite-valued Łukasiewicz logic if it evaluates to 1 under any valuation of propositional variables by real numbers in the interval [0, 1].

Finite-valued and Countable-valued Semantics

Using exactly the same valuation formulas as for real-valued semantics Łukasiewicz also defined (up to isomorphism) semantics over:

- Any finite set of cardinality $n \geq 2$ by choosing the domain as { 0, 1/(n – 1), 2/(n – 1), ..., 1 }.
- Any countable set by choosing the domain as { $p/q \mid 0 \leq p \leq q$ where p is a non-negative integer and q is a positive integer }.

General Algebraic Semantics

The standard real-valued semantics determined by the Łukasiewicz t-norm is not the only possible semantics of Łukasiewicz logic. General algebraic semantics of propositional infinite-valued Łukasiewicz logic is formed by the class of all MV-algebras. The standard real-valued semantics is a special MV-algebra, called the *standard MV-algebra.*

Like other t-norm fuzzy logics, propositional infinite-valued Łukasiewicz logic enjoys completeness with respect to the class of all algebras for which the logic is sound (that is, MV-algebras) as well as with respect to only linear ones. This is expressed by the general, linear, and standard completeness theorems:

The following conditions are equivalent:

- **A** is provable in propositional infinite-valued Łukasiewicz logic.
- **A** is valid in all MV-algebras (*general completeness*).
- **A** is valid in all linearly ordered MV-algebras (*linear completeness*).
- **A** is valid in the standard MV-algebra (*standard completeness*).

Font, Rodriguez and Torrens introduced in 1984 the Wajsberg algebra as an alternative model for the infinite-valued Łukasiewicz logic.

A 1940s attempt by Grigore Moisil to provide algebraic semantics for the n-valued Łukasiewicz logic by means of his Łukasiewicz–Moisil (LM) algebra (which Moisil called *Łukasiewicz algebras*) turned out to be an incorrect model for $n \geq 5$. This issue was made public by Alan Rose in 1956. C. C. Chang's MV-algebra, which is a model for the $\aleph_0$-valued (infinitely-many-valued) Łukasiewicz-Tarski logic, was published in 1958. For the axiomatically more complicated (finite) n-valued Łukasiewicz logics, suitable algebras were published in 1977 by Revaz Grigolia and called MV_n-algebras. MV_n-algebras are a subclass of LM_n-algebras, and the inclusion is strict for $n \geq 5$. In 1982 Roberto Cignoli published some additional constraints that added to LM_n-algebras produce proper models for n-valued Łukasiewicz logic; Cignoli called his discovery *proper Łukasiewicz algebras*.

Gödel–Dummett Logic

Gödel–Dummett logic, also known as Dummett's LC or simply Gödel logic, is another early example of a many-valued logic with truth values in [0,1]. It was introduced by Michael Dummett as the extension of intuitionistic logic by the axiom:

$$(\varphi \to \psi) \vee (\psi \to \varphi).$$

This formula enforces a linear order in the underlying (Kripke-style as well as algebraic) semantics. It also appears in the context of Gödel's observation that it is impossible to characterize intuitionistic logic by finite truth tables. Gödel–Dummett logic can alternatively be obtained as an axiomatic extension of MTL by adding the axiom $\varphi \to \varphi \& \varphi$, which amounts to requiring the idempotence of &, and hence making the interpretation of both conjunctions coincide. In the fuzzy logic setting the Gödel–Dummett logic can be seen as the consequence relation given by the minimum t-norm. It is distinguished as the only t-norm based logic where the truth of a formula in a given evaluation does not depend on the specific values assigned to the propositional variables, but only on the relative order of these values. In this sense, Gödel–Dummett logic can be seen as a logic of comparative truth. Like for Łukasiewicz logic, the computational complexity of testing validity remains coNP-complete.

PREDICATE FUZZY LOGIC

Given any propositional fuzzy logic L there is a uniform way to introduce its first-order counterpart $L\forall$ in a predicate language $\mathcal{PL}$ (defined as in the classical case).

The semantics is given by structures in which predicate symbols are interpreted as functions mapping tuples of domain elements into truth values. More precisely, a structure M consists of a non-empty domain of elements M, a function $f_M : M^n \rightarrow M$ for each n-ary function symbol $f \in \mathcal{PL}$, and a function $P_M : M^n \rightarrow [0,1]$ for each n-ary predicate symbol $P \in \mathcal{PL}$. Fixing an evaluation v of object variables in M, one defines values of terms $\left(\left\|f(t_1,......,t_n)\right\|_v = f_M\left(\left\|t_1\right\|_v,....,\left\|t_n\right\|_v\right)\right)$ and truth values of atomic formulas $\left(\left\|P(t_1,......,t_n)\right\|_v = P_M\left(\left\|t_1\right\|_v,....,\left\|t_n\right\|_v\right)\right)$. Truth values of a universally/existentially quantified formula are computed as infimum/supremum of truth values of instances of the formula where the quantified variable runs over all elements of the domain M. Formally:

$$\left\|(\forall x)\varphi\right\|_v = \inf\left\{\left\|\varphi\right\|_{v[x:a]} \mid a \in M\right\}$$

$$\left\|(\exists x)\varphi\right\|_v = \sup\left\{\left\|\varphi\right\|_{v[x:a]} \mid a \in M\right\},$$

where v[x : a] is the evaluation sending x to a and keeping values of other variables unchanged. The values of other formulas are computed using the truth functions for the propositional connectives of L.

The first-order logic $L\forall$ is then defined as the consequence relation given by preservation of total truth (value 1), as in the propositional case. More precisely, we say that a first-order formula φ is a consequence of a set of formulas Γ (in symbols: $\Gamma \models_{L\forall} \varphi$) if $\left\|\varphi\right\|_v = 1$ for each evaluation v, whenever $\left\|\psi\right\|_v = 1$ for each evaluation v and each $\psi \in \Gamma$.

$L\forall$ can be given a Hilbert-style calculus with the following axioms:

- (P) The (first-order) instances of the axioms of the propositional logic L.
- $(\forall 1)$ $(\forall x)\varphi(x) \rightarrow \varphi t$, where the term t is substitutable for x in.
- $(\exists 1)$ $\varphi(t) \rightarrow (\exists x)\varphi(x)$, where the term t is substitutable for x in.

- $(\forall 2)(\forall x)(\chi \rightarrow \varphi) \rightarrow (\chi \rightarrow (\forall (x)\varphi))$, where x is not free in χ.
- $(\exists 2)(\forall x)(\varphi \rightarrow \chi) \rightarrow ((\exists x)\varphi \rightarrow \chi)$, where x is not free in χ.
- $(\forall 3)(\forall x)(\chi \vee \varphi) \rightarrow \chi \vee (\forall x)\varphi$, where x is not free in χ.

The deduction rules of $L\forall$ are those of L plus the rule of generalization: from φ infer $(\forall x)\varphi$.

For many notable propositional fuzzy logics (including MTL and Gödel logic) the above axiomatic system is sound and complete with respect to the semantics (i.e., $\Gamma \models_{L\forall} \varphi$ iff $\Gamma \vdash_{L\forall} \varphi$ for each Γ and each φ).

However, the first-order Łukasiewicz logic is not recursively axiomatizable as shown by Scarpellini proved that the set of tautologies is actually Σ_2-complete in the sense of arithmetical hierarchy). Completeness can be achieved either by including an infinitary inference rule or by generalizing the set of truth-values. The situation is even more complicated in the case of Hájek's Basic Fuzzy Logic, where the set of first-order tautologies of all structures given by continuous t-norms is as complex as true arithmetics.

Processes in Fuzzy Logic

CHAPTER 4

Fuzzification, negation criterion, minimum criterion, maximum criterion, fuzzy If-Then rules and defuzzification are a few processes that fall under the domain of fuzzy logic. This chapter closely examines these processes used in fuzzy logic to provide an extensive understanding of the subject.

FUZZIFICATION

Fuzzification is the process of changing a real scalar value into a fuzzy value. This is achieved with the different types of fuzzifiers (membership functions). Fuzzy Linguistic Variables are used to represent qualities spanning a particular spectrum.

Temp: {Freezing, Cool, Warm, Hot}

Question: What is the temperature?

Answer: It is warm.

Question: How warm is it?

Temp: {Freezing, Cool, Warm, Hot}

Degree of Truth or "Membership".

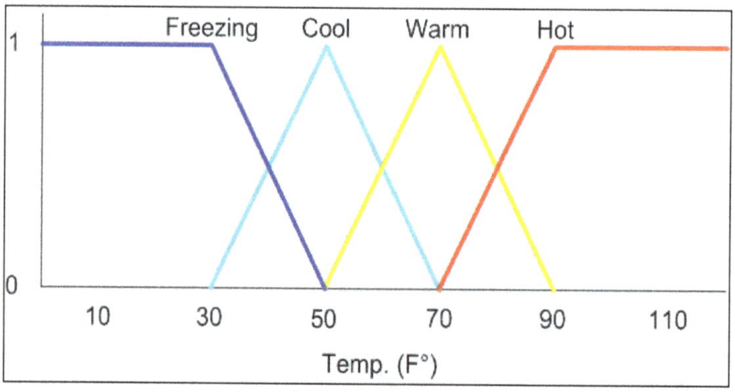

Membership Functions.

How cool is 36 F°?

It is 30% Cool and 70% Freezing.

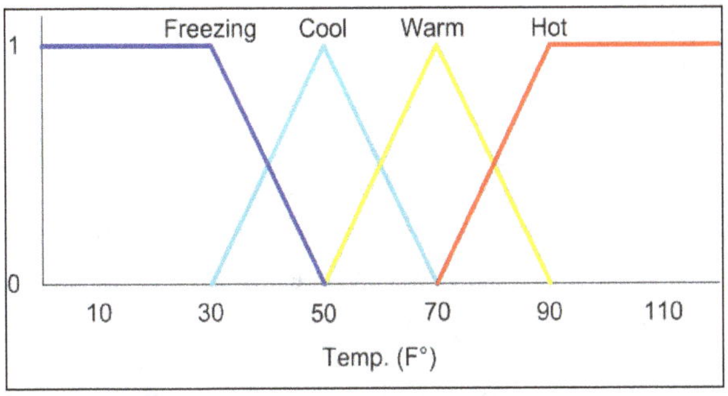

Membership Functions

The MATLAB toolbox includes 11 built-in membership function types. These 11 functions are, in turn, built from several basic functions:

- Piecewise linear functions.
- The Gaussian distribution function.
- The sigmoid curve.
- Quadratic and cubic polynomial curves.

The simplest membership functions are formed using straight lines. The simplest is the triangular membership function, and it has the function name trimf.

The trapezoidal membership function, trapmf, has a flat top and really is just a truncated triangle curve. These straight line membership functions have the advantage of simplicity.

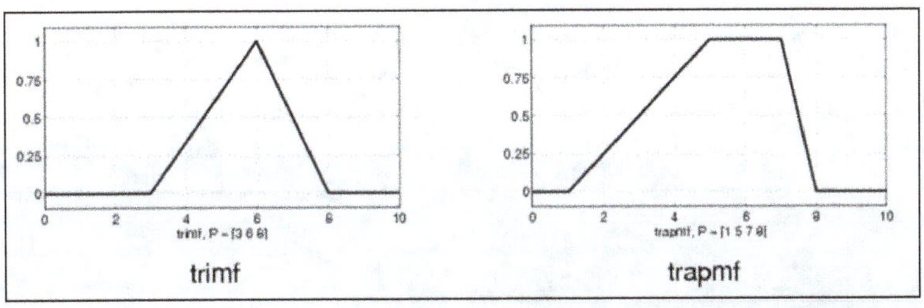

Two membership functions are built on the Gaussian distribution curve: a simple Gaussian curve and a two-sided composite of two different Gaussian curves. The two functions are gaussmf and gauss2mf. The generalized bell membership has the function name gbellmf.

Because of their smoothness and concise notation, Gaussian and bell membership functions are popular methods for specifying fuzzy sets. Both of these curves have the advantage of being smooth and nonzero at all points.

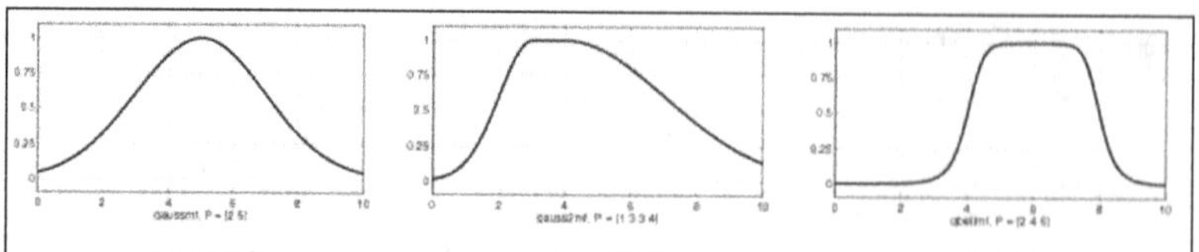

gaussmf gauss2mf gbellmf

Although the Gaussian membership functions and bell membership functions achieve smoothness, they are unable to specify asymmetric membership functions, which are important in certain applications.

The sigmoidal membership function is defined, which is either open left or right. Asymmetric and closed (i.e. not open to the left or right) membership functions can be synthesized using two sigmoidal functions, so in addition to the basic sigmf, you also have the difference between two sigmoidal functions, dsigmf, and the product of two sigmoidal functions psigmf.

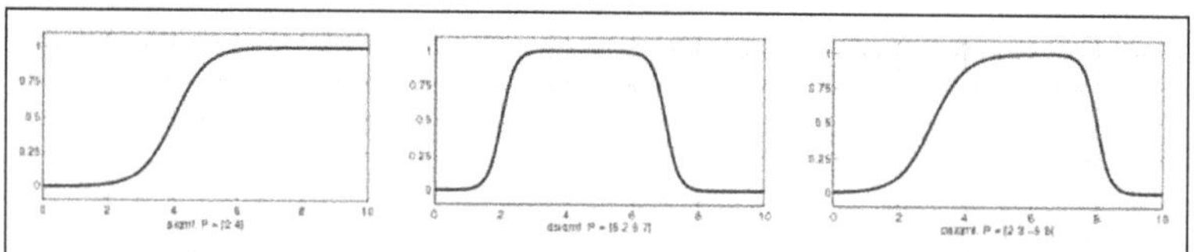

sigmf dsigmf psigmf

Polynomial based curves account for several of the membership functions in the toolbox.

Three related membership functions are the Z, S, and Pi curves, all named because of their shape. The function zmf is the asymmetrical polynomial curve open to the left, smf is the mirror-image function that opens to the right, and pimf is zero on both extremes with a rise in the middle.

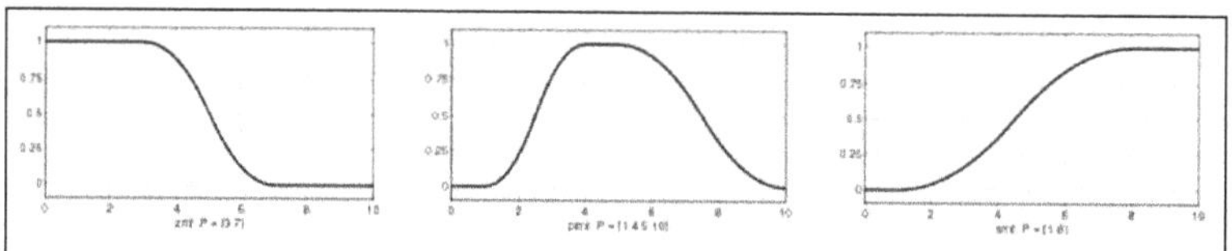

zmf pimf smf

FUZZY LOGIC OPERATIONS

The standard definitions in fuzzy logic as suggested by Lotfi are:

- Negate(negation criterion) : truth (not x) = 1.0 - truth (x).
- Intersection(minimum criterion): truth (x and y) = minimum (truth(x), truth(y)).
- Union(maximum criterion): truth (x or y) = maximum (truth(x), truth(y)).

In other words, consider two fuzzy sets ***A*** and ***B***, the above operations can be defined as following:

- The negate or complement of ***A*** is,

  $$\mu_A^c(\mathbf{x}) = \mathbf{1} - \mu_A(\mathbf{x})$$

- The intersection of A and B,

  $$\mu_{A \cap B}(\mathbf{x}) = \mathbf{min}(\mu_A(\mathbf{x}), \mu_B(\mathbf{x}))$$

- The union of A and B,

  $$\mu_{A \cup B}(\mathbf{x}) = \mathbf{max}(\mu_A(\mathbf{x}), \mu_B(\mathbf{x}))$$

In order to clarify this, a few examples are given. Let **A** be a fuzzy interval between 5 and 8 and **B** be a fuzzy number about 4. The corresponding figures are shown below.

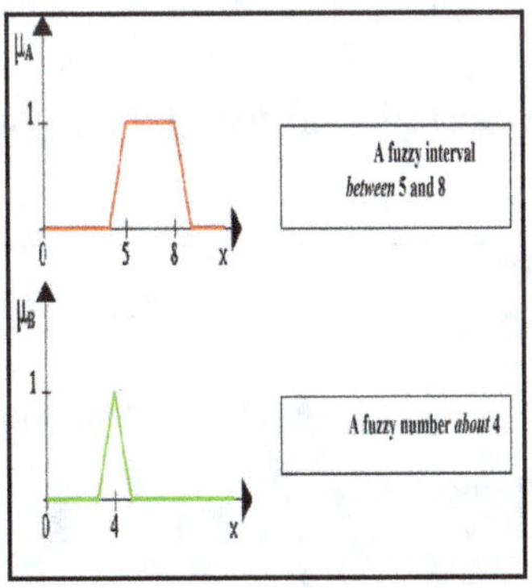

Figure gives an example for a negation. The dotted line is the **NEGATION** of the fuzzy set **A**. To negate the fuzzy set **A**, subtract the membership value in the fuzzy set from one. For example, the membership value at 5 is one. In the negation, the membership value at 5 would be zero (1–1 = 0). If the membership value is 0.4, in the negation, the membership value would be 0.6 (1- 0.4 = 0.6).

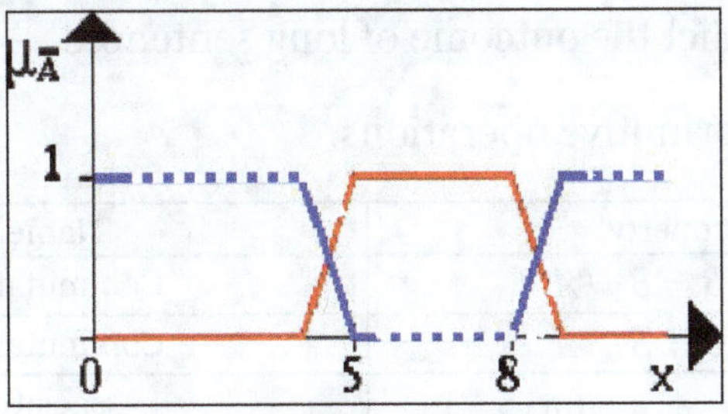

An example for a negation.

Figure shows the fuzzy set between 5 and 8 **AND** about 4. This time the minimum criterion is used, the dotted line represents the result. To find the intersection of these sets take the minimum of the two membership values at each point on the x-axis. For example, in the figure **A** fuzzy set has a membership of ZERO when x = 4 and **B** fuzzy set has a membership of ONE when x = 4. The intersection would have a membership value of ZERO when x = 4 because the minimum of zero and one is zero.

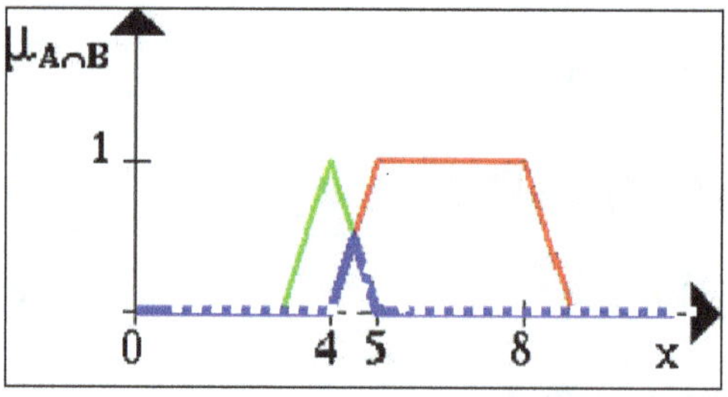

The fuzzy set between 5 and 8 AND about 4.

Finally, the Fuzzy set between 5 and 8 **OR** about 4 is shown in figure. This time the maximum criterion is used. To find the union of these sets take the maximum of the two membership values at each point on the x-axis. For example, ***A*** fuzzy set has a membership of ZERO when x = 4 and **B** fuzzy set has a membership of ONE when x = 4. The union would have a membership value of ONE when x = 4 because the maximum of zero and one is one.

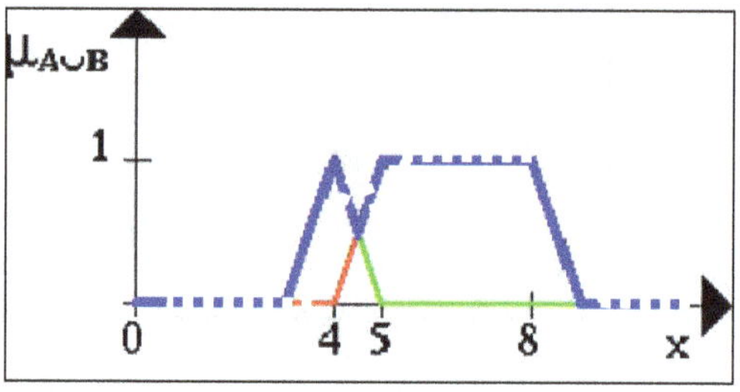

The Fuzzy set between 5 and 8 OR about 4.

These basic operations, provide guidelines to construct more complex ones which in turn can be used to create fuzzy machines.

The operations $\cup$ and $\cap$ associate, commute and more. These properties are important, because they help to predict the outcome of long sentences.

Table: Properties of the primitive operations.

Property	Name
$A \cup B = B \cup A$	Commutative
$A \cap B = B \cap A$	Commutative
$(A \cup B) \cup C = A \cup (B \cup C)$	Associative
$(A \cap B) \cap C = A \cap (B \cap C)$	Associative
$A \cap (B \cup C) = (A \cap B) \cup (A \cap C)$	Distributive

$A\cup(B\cap C)=(A\cup B)\cup(A\cup C)$	Distributive
$\overline{A\cap B}=\overline{A}\cup\overline{B}$	DeMorgan
$\overline{A\cup B}=\overline{A}\cap\overline{B}$	DeMorgan
$(A\cap B)\cup A=A$	Absorption
$(A\cup B)\cap A=A$	Absorption
$A\cup A=A$	Idempotency
$A\cap A=A$	Idempotency
$A\cup\overline{A}\neq 1$	Exclusion not satisfied
$A\cap\overline{A}\neq 0$	Exclusion not satisfied

There are some operators in the fuzzy expression such as $\neg$ for (negation), $\wedge$ for (conjunction), $\vee$ for (disjunction), and $\rightarrow$ for (implication).

Negation "not":	$\neg a=1-a$
Conjunction "and" "$\wedge$":	$a\wedge b=\text{Min}(a,\ b)$
Disjunction "or" "$\vee$":	$a\vee b=\text{Max}(a,\ b)$
Implication "$\rightarrow$":	$a\rightarrow b=\text{Min}(1,\ 1+b-a)$

The following special symbols can also be used to represent fuzzy logic operations and connective structure:

$\Rightarrow$ for "if – then"

$\Leftrightarrow$ for "if and only if"

The union of **A** and **B** is then a fuzzy set $\mu_{A\cup B}$ given by,

$$\mu_{A\cup B}(\mathbf{x})=\mathbf{max}\{\mu_A(\mathbf{x}),\mu_B(\mathbf{x})\}$$

Similarly, the intersection $\mu_{A\cap B}(x)$ is given by,

$$\mu_{A\cap B}(\mathbf{x})=\mathbf{min}\{\mu_A(\mathbf{x}),\mu_B(\mathbf{x})\}$$

And the complement μ_A is defined according to,

$$\mu^{\neg}_{A}=\mathbf{1}-\mu A$$

Example: Assume a in the [0, 1]. Prove that $a\wedge\neg a\neq 0$.

$$a\wedge\neg a=\text{Min}(a,\neg a)=\text{Min}(a,1-a)$$

$$=\begin{cases} a & \text{if } 0\le a\le 0.5 \\ 1-a & \text{if } 0.5\le a<1 \end{cases}$$

$$\therefore 0<a\wedge\neg a\le 0.5$$

Then,

$$a \wedge \neg a \neq 0$$

Example: Assume a in the [0, 1] then,

$$a \vee \neg a = \text{Max}(a, \neg a) = \text{Max}(a, 1-a)$$

$$= \begin{cases} a & \text{if } 0.5 \leq a < 1 \\ 1-a & \text{if } 0 < a \leq 0.5 \end{cases}$$

$$\therefore a \vee \neg a = 1 \text{ if } a = 0 \text{ or } a = 1$$

$$a \vee \neg a < 1 \quad \text{otherwise}.$$

Example: Let a = 1, and b = 0 then,

$$(1)\bar{a} = 0$$
$$(2) a \wedge b = \text{Min}(1,0) = 0$$
$$(3) a \vee b = \text{Max}(1,0) = 1$$
$$(4) a \rightarrow b = \text{Min}(1, 1-1+0) = 0$$

Example: Let a = 1, and b = 1 then,

$$(1)\bar{a} = 0$$
$$(2) a \wedge b = \text{Min}(1,1) = 1$$
$$(3) a \vee b = \text{Max}(1,1) = 1$$
$$(4) a \rightarrow b = \text{Min}(1, 1-1+1) = 1$$

Example: Let a = 0.6, and b = 0.7 then,

$$(1)\bar{a} = 0.4$$
$$(2) a \wedge b = \text{Min}(0.6, 0.7) = 0.6$$
$$(3) a \vee b = \text{Max}(0.6, 0.7) = 0.7$$
$$(4) a \rightarrow b = \text{Min}(1, 1-0.6+0.7) = \text{Min}(1, 1.1) = 1$$

Basic Concepts Associated with Fuzzy Set

- The support of a fuzzy set **A** in the universe of discourse **U** is a crisp set that contain all elements of **U** that have nonzero value in **A**, that is,

$$S(A) = \{x \in U \mid \mu_A(x) > 0\}$$

- The crossover point of a fuzzy set is the point in ***U*** whose membership value in ***A*** equals 0.5.
- An α – ***cut*** *or* α – ***level set*** of a fuzzy set ***A*** is a crisp set A_α is that contains all elements in ***U*** that have the membership values in ***A*** greater than or equal to α, that is,

  $$A_\alpha = \{x \in U \mid \mu_A(x) \geq \alpha\}$$

  $A_\alpha = \{x \in U \mid \mu_A(x) > \alpha\}$ is called "**strong α -level set**" or "**strong α -cut**"
- The height of a fuzzy set is the largest membership value attained by any point.
- The cardinality or "power" of a fuzzy set ***A*** is defined as,

  $$|A| = \sum_{x \in X} \mu_A(x)$$

And

$$\|A\| = \frac{|A|}{|X|}$$

Obviously the relative cardinality of a fuzzy set depends on the cardinality of the universe. So you have to choose the same universe if you want to compare fuzzy sets by their relative cardinality.

Example**:** A realtor wants to classify the house he offers to his clients.

One indicator of comfort of these houses is the number of bedrooms in it. Let X= {1, 2, 3, 4, ..., 10} be the set of available types of houses described by x = number of bedrooms in a house. Then the fuzzy set "comfortable type of house for a 4-person family" may be described as:

$$A = \{(1, .2), (2, .5), (3, .8), (4, 1), (5, .7), (6, .3)\}$$

- The support of S(A) = {1,2, 3, 4, 5, 6}. The elements (types of houses) {7, 8, 9, 10} are not part of the support of A.
- A list of possible α -level sets:

  $$\mathbf{A_{.2} = \{1,2,3,4,5,6\}}$$
  $$\mathbf{A_{.5} = \{2,3,4,5\}}$$
  $$\mathbf{A_{.8} = \{3,4\}}$$
  $$\mathbf{A_{1} = \{4\}}$$

  The strong α -level set for α = .8 is A.8 = {4}.

The cardinality is

$$|A| = .2 + .5 + .8 + 1 + .7 + .3 = 3.5$$

And its relative cardinality is

$$||A|| = 3.5/10 = 0.35$$

Further Operations on Fuzzy Sets

Algebraic Operations

- The Cartesian product of fuzzy sets is defined as follows: Let $A_1, \ldots, A_n$ be fuzzy sets in $X_1, \ldots, X_n$. The Cartesian product is then a fuzzy set in the product space $X_1 \times \ldots \times X_n$ with the membership function,

$$\mu_{(A1\times\ldots\times A_n)}(\mathbf{x}) = \min_{i}\{\mu_{Ai}(\mathbf{x}_i) \mid \mathbf{x} = (\mathbf{x}_1, \ldots.\mathbf{x}_n), \mathbf{x}_i \in X_i\}$$

- The mth power of a fuzzy set A is a fuzzy set with the membership function,

$$\mu_{A^m}(x) = [\mu_A(x)]^m, \qquad x \in X$$

- The algebraic sum (probabilistic sum) **C = A + B** is defined as:

$$C = \{(x, \mu_{A+B}(x)) \mid x \in X\}$$

Where,

$$\mu_{A+B}(x) = \mu_A(x) + \mu_B(x) - \mu_A(x).\mu_B(x)$$

- The bounded sum $C = A \oplus B$ is defined as:

$$C = \{(x, \mu_{A\oplus B}(x)) \mid x \in X\}$$

Where,

$$\mu_{A\oplus B}(x) = \min\{1, \mu_A(x) + \mu_B(x)\}$$

- The bounded difference $C = A \ominus B$ is defined as:

$$C = \{(x, \mu_{A\ominus B}(x)) \mid x \in X\}$$

Where,

$$\mu_{A\ominus B}(x) = \max\{(0, \mu(x) + \mu_B(x) - 1)\}$$

- The algebraic product of two fuzzy sets $C = \mathrm{A} \bullet \mathrm{B}$ is defined as

$$C = \{(x, \mu_A(x).\mu B(x)) \mid x \in X\}$$

Example:

$$Let \quad A(x) - \{(3,.5),(5,1),(7,.6)\}$$
$$B(x) = \{(3,1),(5,.6)\}$$

The above definitions are then illustrated by the following rules:

$$A \times B = \{[(3;3),.5][(5;3),1],[(7;3),.6],$$
$$[(3;5),.5],[(5;5),.6],[(7;5),6]\}$$
$$A^2 = \{(3,.25),(5,1),(7,.36)\}$$

$$A + B = \{(3,1),(5,1),(7,6)\}$$
$$A \oplus B = \{(3,1),(5,1),(5,.6)\}$$
$$A \odot B = \{(3,.5),(5,.6)\}$$

Set-theoretic Operations

The intersection of fuzzy sets, interpreted as the logical "and," was modeled as the min-operator and the union, interpreted as "or," as the max-operator. Other operators have also been suggested. We shall investigate one basic class of operators: operators for the intersection and union of fuzzy sets—referred to as triangular norms and conorms.

T-norms

T-norms are two-valued functions from [0,1] × [0,1] which satisfy the following conditions:

- $t(0,\ 0) = 0;\ t(\mu_A(x),\ 1) = t(1, \mu_A(x)) = \mu_A(x),\ x \in X$ unit element
- $t(\mu_A(x), \mu_B(x)) \le t(\mu_C(x), \mu_D(x))$

  if $\mu_A(x) \le \mu_C(x)$ and $\mu_B(x) \le \mu_D(x)$ (monotonicity)
- $t(\mu_A(x), \mu_B(x)) = t(\mu_B(x), \mu_A(x))$ (commutativity)
- $t(\mu_A(x), t(\mu_B(x), \mu_C(x))) = t(t(\mu_A(x), \mu_B(x)), \mu_C(x))$ (associatitivty)

The functions t define a general class of intersection operators for fuzzy sets. The operators belonging to this class of t-norms are, in particular, associative and therefore it is possible to compute the membership values for the intersection of more than two fuzzy sets by recursively applying a t-norm operator.

T-conorms (or S-norms)

t-conorms or s-norms are associative, commutative, and monotonic two placed functions s, which map from [0, 1] × [0, 1] into [0, 1]. These properties are formulated with the following conditions:

- $s(1,1)=1;\ s(\mu_A(x),\ 0)=s(0,\ \mu_A(x))=\mu_A(x),\ x\in X$ unit element
- $s(\mu_A(x),\mu_B(x))\le s(\mu_C(x),\ \mu_D(x))$
  if $\mu_A(x)\le\mu_C(x)$ and $\mu_B(x)\le\mu_D(x)$ (monotonicity)
- $s(\mu_A(x),\mu_B(x))=s(\mu_B(x),\mu_A(x))$ (commutativity)
- $s(\mu_A(x),\ s(\mu_B(x),\ \mu_C(x)))\ =\ s(s(\mu_A(x),\ \mu_B(x)),\ \mu_C(x))$ (associatitivty)

Alsina defined the function t as:

$$t(\mu_A(x),\ \mu_B(x))=1-s(1-\mu_A(x),\ 1-\mu_B(x))$$

So any t-conorm s can be generated from a t-norm t through this transformation. More general, for suitable negation operators like the complement operator for fuzzy sets, defined as $n(\mu_A(x))=(1-\mu_A(x))$, pairs of t-norms ***t*** and t-conorms **s** satisfy the following generalization of DcMorgans law:

$$s(\mu_A(x),\ \mu_B(x))=n(t(n(\mu_A(x)),\ n(\mu_B(x)))),\qquad \text{and}$$

$$t(\mu_A(x),\ \mu_B(x))=n(s(n(\mu_A(x)),\ n(\mu_B(x))))\qquad x\in X$$

FUZZY IF-THEN RULES

Fuzzy sets and fuzzy sets operations are the subjects and verbs of fuzzy logic. If-Then rule statements are used to formulate the conditional statements that comprise fuzzy logic.

A single fuzzy If-Then rule assumes the form

If **x** is $\boldsymbol{A_1}$ Then **y** is $\boldsymbol{B_2}$

where A_1 and B_2 are linguistic variables defined by fuzzy sets on the ranges (i.e. universe of discourse) X and Y respectively. The If-part of the rule 'x is A_1' is called the antecedent or premise and the Then-part of the rule 'y is B_2' is called the consequent.

In other words, the conditional statement can be expressed in a mathematical form

$$\text{If } A_1 \text{ Then } B_2 \text{ or } A_1 \rightarrow B_2$$

Example:

Speed and pressure of a steam engine can be expressed with the following linguistic conditional statement

If **Speed** is **Slow** Then **Pressure** should be **High**.

Graphically, this statement looks like:

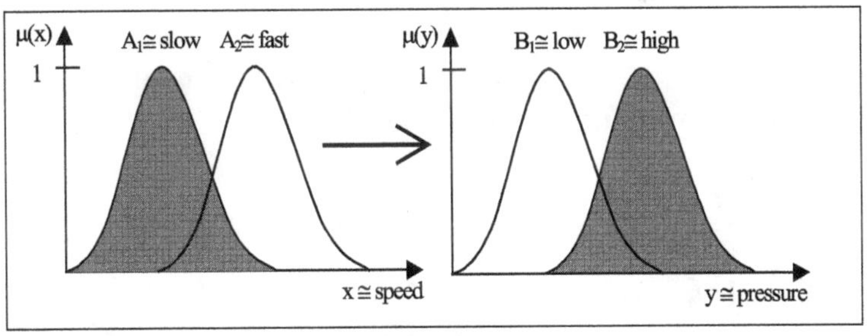

If-Then rule.

Rule Forms

In general, three forms of rule exist for any linguistic variables.

- Assignment statement:

  e.g. x is not large AND not very small.

- Conditional statement:

  e.g. IF x is very big THEN y is medium.

- Unconditional statement:

  e.g. set pressure high.

Compound Rules

A linguistic statement expressed by a human might involve compound rule structures. By using basic properties and operations defined for fuzzy sets, any compound rule structure may be decomposed and reduced to a number of simple canonical rules.

Conjunctive antecedents: For a multiple conjunctive antecedents of the form:

$$\text{IF x is } A^1 \text{ AND x is } A^2 \ldots \text{ AND x is } A^n \text{ THEN y is } B^s$$

Can be rewritten as,

IF x is A^s THEN y is B^s

Where $A^s = A^1 \cap A^2 \cap A^3 \cap \ldots \cap A^n$ and A^s is expressed by means of membership function based on the definition of fuzzy intersection operation as:

$$\mu_{A^s}(x) = \min\left\lfloor \mu_{A^1}(x), \mu_{A^2}(x), \ldots, \mu_{A^n}(x) \right\rfloor$$

Disjunctive antecedents: For a multiple disjunctive antecedents of the form:

IF x is A^1 OR x is A^2 ... OR x is A^n THEN y is B^s

Can be rewritten as,

IF x is A^s THEN y is B^s

Where $A^s = A^1 \cup A^2 \cup A^3 \cup \ldots \cup A^n$ and A^s is expressed by means of membership function based on the definition of fuzzy union operation as:

$$\mu_{A^s}(x) = \max\left\lfloor \mu_{A^1}(x), \mu_{A^2}(x), \ldots, \mu_{A^n}(x) \right\rfloor$$

Aggregation of Rules

Most rule-based systems have more than one rule. The process of obtaining the overall of the consequent from the individual consequent contributed by each rule in the rule-base is the aggregation of rules.

In the case of a system of rules that must be jointly satisfied, the rules are connected by AND connectives. The aggregated output y is found by fuzzy intersection of the entire individual rule consequent y^i, where $i = 1, 2, 3, \cdots, r$.

y = y^1 AND y^2 AND ... AND y^r

or

$$y = y^1 \cap y^2 \cap \ldots \cap y^r$$

Which is defined by the membership function:

$$\mu_y(y) = \min\left\lfloor \mu_{y^1}(y), \mu_{y^2}(y), \ldots, \mu_{y^1}(y) \right\rfloor \text{ for } y \in Y$$

For the case of a disjunctive system of rules where at least one rule must be satisfied, the rules are connected by OR connectives. The aggregated output y is found by fuzzy union of all the individual rule consequent y^i, where $i = 1, 2, 3, \cdots, r$.

y = y^1 OR y^2 OR ... OR y^r

or

$$y = y^1 \cup y^2 \cup \ldots \cup y^r$$

Which is defined by the membership function,

$$\mu_y(y) = \max\left[\mu_{y^1}(y), \mu_{y^2}(y), \ldots, \mu_{y^r}(y)\right] \text{ for } y \in Y$$

Example:

Lets consider a fuzzy system with two inputs x_1 and x_2 (antecedents) and a single output y (consequent). Inputs x_1 and x_2 have three linguistic variables small, medium and big with triangular membership function. Output y has two linguistic variables small and big with triangular membership function. The rule-base consists of the two following rules:

IF x_1 is small AND x_2 is medium THEN y is big

IF x_1 is medium AND x_2 is big THEN y is small

Inputs x_1 and x_2 are crisp values and the membership values of inputs x_1 and x_2 are calculated for triangular membership functions. The aggregated output for r rules is given by

$$\mu(y) = \max^i\left[\min\left[\mu(x_1^i), \mu(x_2^i)\right]\right]$$

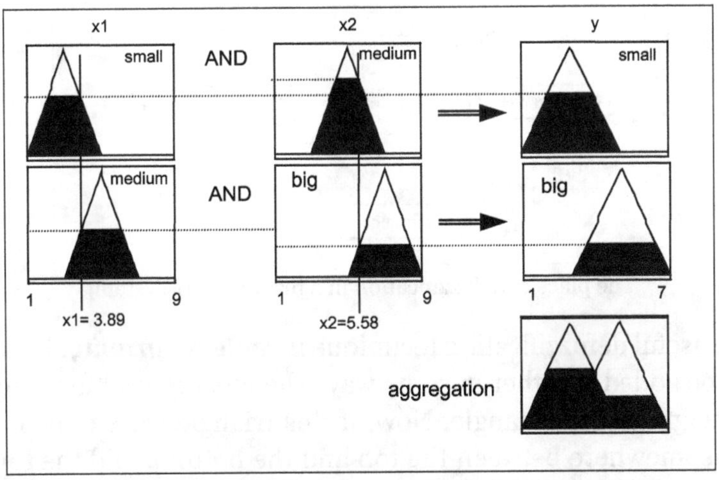

Max-min inference method.

Where $i = 1,2,3,\ldots.r$. In this example r equals to 2. The minimum membership value for the antecedents propagates through to the consequent and truncates the membership function for the consequent of each rule. The truncated membership functions for each rule are aggregated using the graphical equivalent of either for conjunctive or disjunctive rules.

The aggregation operation max results in an aggregated membership function comprised of the outer envelope of the individual truncated membership forms from each rule. Figure below illustrates the rule-base and aggregation process using max-min method of inference.

DEFUZZIFICATION

Defuzzification is the process of producing a quantifiable result in Crisp logic, given fuzzy sets and corresponding membership degrees. It is the process that maps a fuzzy set to a crisp set. It is typically needed in fuzzy control systems. These will have a number of rules that transform a number of variables into a fuzzy result, that is, the result is described in terms of membership in fuzzy sets. For example, rules designed to decide how much pressure to apply might result in "Decrease Pressure (15%), Maintain Pressure (34%), Increase Pressure (72%)". Defuzzification is interpreting the membership degrees of the fuzzy sets into a specific decision or real value.

The simplest but least useful defuzzification method is to choose the set with the highest membership, in this case, "Increase Pressure" since it has a 72% membership, and ignore the others, and convert this 72% to some number. The problem with this approach is that it loses information. The rules that called for decreasing or maintaining pressure might as well have not been there in this case.

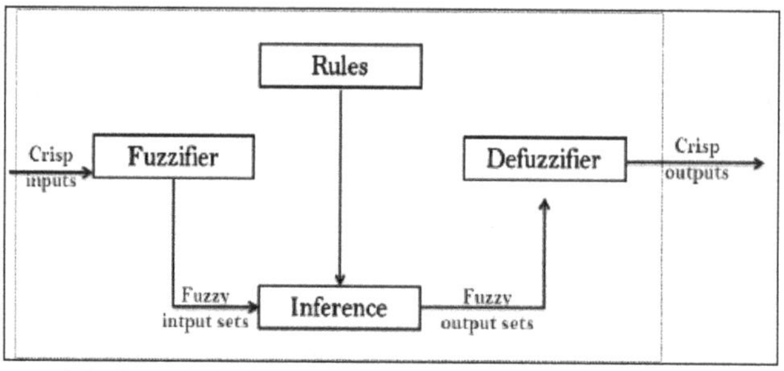

The place of defuzzification in a fuzzy control system.

A common and useful defuzzification technique is *center of gravity*. First, the results of the rules must be added together in some way. The most typical fuzzy set membership function has the graph of a triangle. Now, if this triangle were to be cut in a straight horizontal line somewhere between the top and the bottom, and the top portion were to be removed, the remaining portion forms a trapezoid. The first step of defuzzification typically "chops off" parts of the graphs to form trapezoids (or other shapes if the initial shapes were not triangles). For example, if the output has "Decrease Pressure (15%)", then this triangle will be cut 15% the way up from the bottom. In the most common technique, all of these trapezoids are then superimposed one upon another, forming a single geometric shape. Then, the centroid of this shape, called the *fuzzy centroid*, is calculated. The x coordinate of the centroid is the defuzzified value.

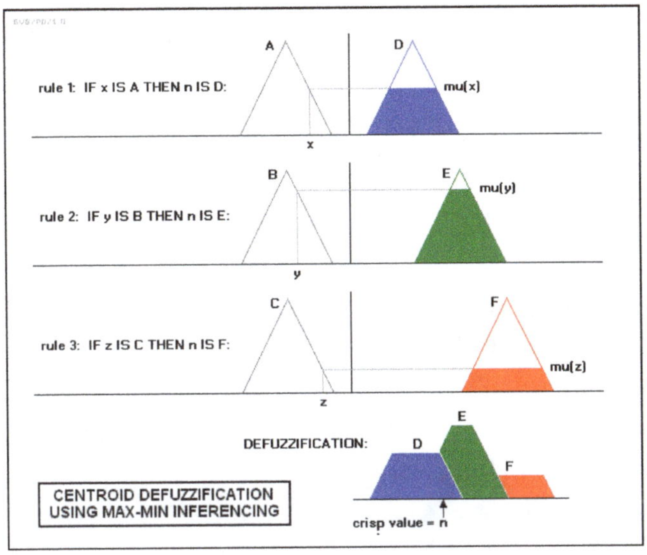

A particular defuzzification method.

Defuzzification Methods

Fuzzy rule based systems evaluate linguistic if-then rules using fuzzification, inference and composition procedures. They produce fuzzy results which usually have to be converted into crisp output. To transform the fuzzy results in to crisp, defuzzification is performed.

Defuzzification is the process of converting a fuzzified output into a single crisp value with respect to a fuzzy set. The defuzzified value in FLC (Fuzzy Logic Controller) represents the action to be taken in controlling the process.

Different Defuzzification Methods

The following are the known methods of defuzzification.

- Center of Sums Method (COS).
- Center of gravity (COG) / Centroid of Area (COA) Method.
- Center of Area / Bisector of Area Method (BOA).
- Weighted Average Method.
- Maxima Methods:
    - First of Maxima Method (FOM).
    - Last of Maxima Method (LOM).
    - Mean of Maxima Method (MOM).

Center of Sums (COS) Method

This is the most commonly used defuzzification technique. In this method, the overlapping area is counted twice.

The defuzzified value x^* is defined as :

$$x^* = \frac{\Sigma_{i=1}^{N} x_i \cdot \Sigma_{k=1}^{n} \mu_{A_k}(x_i)}{\Sigma_{i=1}^{N} \quad \Sigma_{k=1}^{n} \mu_{A_k}(x_i)},$$

Here, n is the number of fuzzy sets, N is the number of fuzzy variables, $\mu_{A_k}(x_i)$ is the membership function for the k-th fuzzy set.

Example:

The defuzzified value x^* is defined as:

$$x^* = \frac{\Sigma_{i=1}^{N} A_i \times \overline{x_i}}{\Sigma_{i=1}^{N} A_i},$$

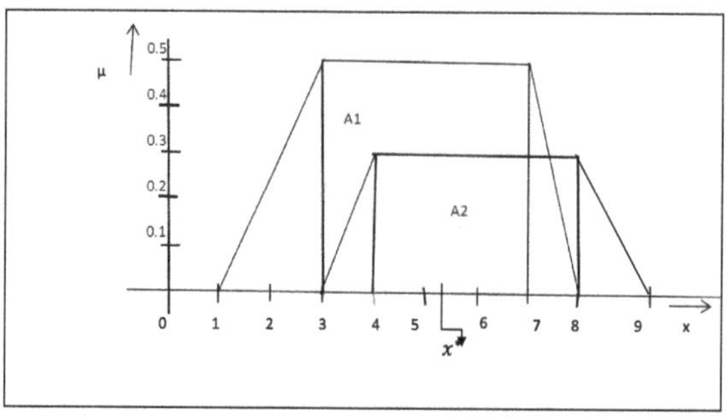

Fuzzy sets C_1 and C_2.

Here, A_i represents the firing area of ith rules and k is the total number of rules fired and $\overline{x_i}$ represents the center of area.

The aggregated fuzzy set of two fuzzy sets C_1 and C_2 is shown in figure. Let the area of this two fuzzy sets are A_1 and A_2.

$$A_1 = \frac{1}{2} * \left[(8-1)+(7-3)\right] * 0.5 = \frac{1}{2} * 11 * 0.5 = 55/20 = 2.75$$

$$A_2 = \frac{1}{2} * \left[(9-3)+(8-4)\right] * 0.3 = \frac{1}{2} * 10 * 0.3 = 3/2 = 1.5$$

Now the center of area of the fuzzy set C_1 is let say $\overline{x}_1 = (7+3)/2 = 5$ and the center of area of the fuzzy set C_2 is $\overline{x}_2 = (8+4)/2 = 6$.

Now the defuzzified value $x^* = \frac{\left(A_1.\overline{x_1} + A_2.\overline{x_2}\right)}{A_1 + A_2} = \frac{(2.75*5+1.5*6)}{(2.75+1.5)} = 22.75/4.25 = 5.35$

Center of gravity (COG) / Centroid of Area (COA) Method

This method provides a crisp value based on the center of gravity of the fuzzy set. The total area of the membership function distribution used to represent the combined control action is divided into a number of sub-areas. The area and the center of gravity or centroid of each sub-area is calculated and then the summation of all these sub-areas is taken to find the defuzzified value for a discrete fuzzy set.

For discrete membership function, the defuzzified value denoted as x^* using COG is defined as:

$x^* = \frac{\sum_{i=1}^{n} x_i \cdot \mu(x_i)}{\sum_{i=1}^{n} \mu(x_i)}$, Here x_i indicates the sample element, $\mu(x_i)$ is the membership function, and n represents the number of elements in the sample.

For continuous membership function, x^* is defined as:

$$x^* = \frac{\int x\,\mu_A(\mathrm{x})\,\mathrm{dx}}{\int \mu_A(\mathrm{x})\,\mathrm{dx}}$$

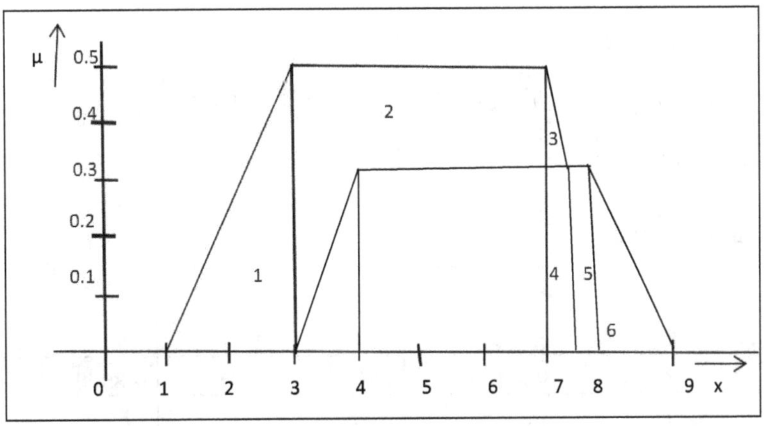

Fuzzy sets C_1 and C_2.

Example:

The defuzzified value x^* using COG is defined as:

$x^* = \frac{\sum_{i=1}^{N} A_i \times \overline{x_1}}{\sum_{i=1}^{N} A_i}$, Here N indicates the number of sub-areas, A_i and $\overline{x_1}$ represents the area and centroid of area, respectively, of ith sub-area.

In the aggregated fuzzy set as shown in figure., the total area is divided into six sub-areas.

For COG method, we have to calculate the area and centroid of area of each sub-area.

These can be calculated as below:

- The total area of the sub-area 1 is ½ * 2 * 0.5 = 0.5.
- The total area of the sub-area 2 is (7-3) * 0.5 = 4 * 0.5 = 2.
- The total area of the sub-area 3 is ½ * (7.5-7) * 0.2 = 0.5 * 0.5 *0.2 =.05.
- The total area of the sub-area 4 is 0.5* 0.3 = .15.
- The total area of the sub-area 5 is 0.5* 0.3 = .15.
- The total area of the sub-area 6 is ½ *1* 0.3 = .15.

Now the centroid or center of gravity of these sub-areas can be calculated as:

- Centroid of sub-area1 will be (1+3+3)/3 = 7/3 =2.333.
- Centroid of sub-area2 will be (7+3)/2 = 10/2 = 5.
- Centroid of sub-area3 will be (7+7+7.5)/3 = 21.5/3 =7.166.
- Centroid of sub-area4 will be (7+7.5)/2 =14.5/2=7.25.
- Centroid of sub-are5 will be (7.5+8)/2 =15.5/2 = 7.75.
- Centroid of sub-area6 will be (8+8+9)/3 = 25/3 = 8.333.

Now we can calculate $A_i \cdot \overline{x}_1$ and is shown in table:

Sub-area number	$Area(A_i)$	Centroid of area $(\overline{\chi}_i)$	$A_i \cdot \overline{x}_1$
1	0.5	2.333	1.1665
2	02	5	10
3	.05	7.166	0.3583
4	.15	7.25	1.0875
5	.15	7.25	1.1625
6	.15	8.333	1.2499

The defuzzified value x^* will be $\dfrac{\Sigma_{i=1}^{N} A_i \times \overline{x}_l}{\Sigma_{i=1}^{N} A_i}$

$$= \frac{(1.1665+10+0.3583+1.0875+1.1625+1.2499)}{(0.5+2+.05+.15+.15+.15)}$$

$$=(15.0247)/3 = 5.008$$

$$x^* = 5.008$$

Center of Area / Bisector of Area Method (BOA)

This method calculates the position under the curve where the areas on both sides are equal.

The BOA generates the action that partitions the area into two regions with the same area.

$\int_{a}^{x^*} \mu_A(x)\, dx = \int_{x^*}^{\beta} \mu_A(x)\, dx$, where α = min {x| x ∈ X} and β = max {x| x ∈ X}

Weighted Average Method

This method is valid for fuzzy sets with symmetrical output membership functions and produces results very close to the COA method. This method is less computationally intensive. Each membership function is weighted by its maximum membership value. The defuzzified value is defined as:

$$x^* = \frac{\Sigma\mu(x).x}{\Sigma\mu(x)}$$

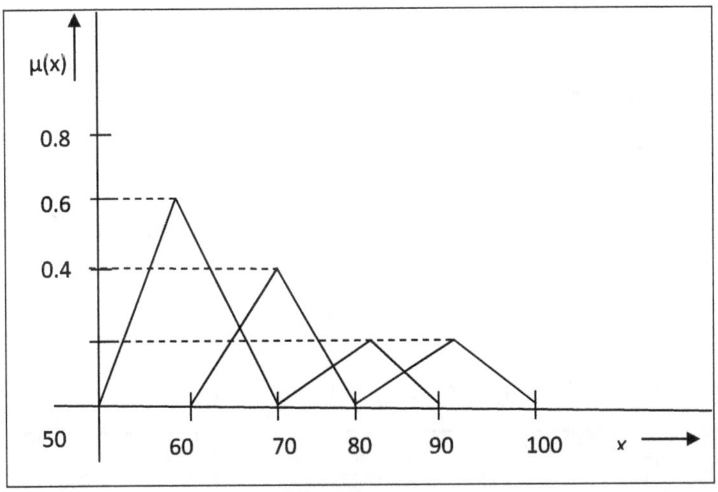

Fuzzy set A.

Here Σ denotes the algebraic summation and x is the element with maximum membership function.

Example:

Let A be a fuzzy set that tells about a student as shown in figure and the elements with corresponding maximum membership values are also given.

A = {(P, 0.6), (F, 0.4),(G, 0.2),(VG, 0.2), (E, 0)}

Here, the linguistic variable P represents a Pass student, F stands for a Fair student, G represents a Good student, VG represents a Very Good student and E for an Excellent student.

Now the defuzzified value x^* for set A will be

$$x^* = \frac{(60*0.6+70*0.4+80*0.2+90*0.2+100*0)}{0.6+0.4+0.2+0.2+0}$$
$$= 98/1.4 = 70$$

The defuzzified value for the fuzzy set A with weighted average method represents a Fair student.

Maxima Methods

This method considers values with maximum membership. There are different maxima methods with different conflict resolution strategies for multiple maxima.

- First of Maxima Method (FOM).
- Last of Maxima Method (LOM).
- Mean of Maxima Method (MOM).

First of Maxima Method (FOM)

This method determines the smallest value of the domain with maximum membership value.

Example:

The defuzzified value x^* of the given fuzzy set will be $x^* = 4$.

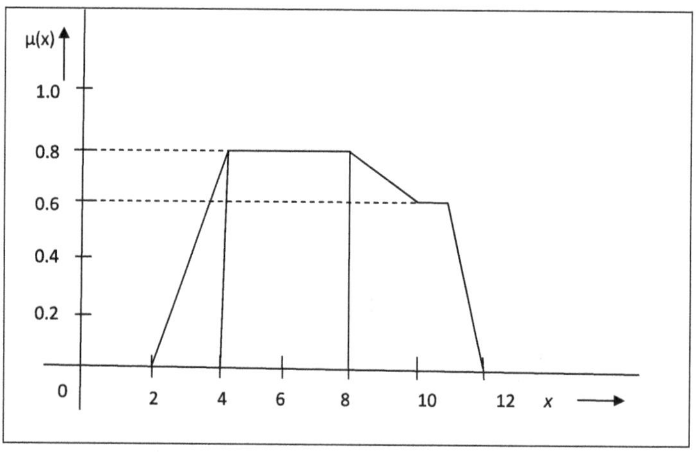

Last of Maxima Method (LOM)

Determine the largest value of the domain with maximum membership value.

In the example given for FOM, the defuzzified value for LOM method will be $x^* = 8$.

Mean of Maxima Method (MOM)

In this method, the defuzzified value is taken as the element with the highest membership values. When there are more than one element having maximum membership values, the mean value of the maxima is taken.

Let A be a fuzzy set with membership function μ_A defined over $x \in X$, where X is a universe of discourse. The defuzzified value is let say x^* of a fuzzy set and is defined as,

$$x^* = \frac{\Sigma_{x_i} \in M \ x_i}{|M|},$$

Here, $M \quad \{x_i \mid \quad _A(x_i)$ is equal to the height of the fuzzy set A} and |M| is the cardinality of the set M.

Example:

In the example as shown in figure. , x = 4, 6, 8 have maximum membership values and hence |M| = 3

According to MOM method, $x^* = \frac{\Sigma x_{i \in M} x_i}{|M|}$

Now the defuzzified value x^* will be $x^* = \frac{4+6+8}{3} = \frac{18}{3} = 6.$

Applications

5 CHAPTER

Fuzzy logic has been used in numerous applications such as facial pattern recognition, air conditioners, washing machines, vacuum cleaners, antiskid braking systems, transmission systems, control of subway systems and unmanned helicopters, robot navigation, etc. All these applications of fuzzy logic have been carefully analyzed in this chapter.

FUZZY LOCATING SYSTEM

Fuzzy locating is a method based on appropriate measuring technology for estimating a location of an object. The term fuzzy relates to rough spatial coincidence or contiguity assessment compared to the alternative crisp locating, which derives precise coordinates of a location of an object. The fuzzy part of the locating process balances the physical and the mathematical portions of processing measurement data of the objects involved and a priori knowledge with the operational ambience.

The result of fuzzy locating shall suffice for operational support and not for metric confirmation of measures taken at earlier occasions. Fuzzy locating compares with the distinction respectively segregation of logics with the terms crisp sets and relations or fuzzy sets and relations.

Systematic and stochastic errors occurring under operational requirements and conditions turn virtually precise measures in a friendly ambience to fuzzy metrics and in worse ambient conditions to erratic results. In consequence the trade off between technical effort and operational support adjusts to physical limitations. Such balancing deliberately neglects classical terms of precision in favour of a strong commitment for operational unambiguity. This balancing leads to the less precise fuzzy locating: As a more general approach, fuzzy locating is a method for best estimating a roughly determined location of an object as a distinction of operational contiguity, for example current presence in a room or current existence on a known place.

For the locating task an object to be located must be at least equipped with a wireless target in a wireless communication environment. Each contributing wireless target operates in a wireless communication environment. Prerequisite for radio frequency based wireless cooperation is some cohesion of the wireless nodes in a networking concept. Each wireless target has at least one physical propagation parameter that varies with location. Better qualified approaches make use of more than one physical parameter. Alternatively optical and acoustical solutions are known.

Variation of parameters is partly deterministic with varying distances between wireless nodes. A location estimate approximating the real location of the receiver preferably under real time constraints is determined on the basis of a statistical model for the process of observation in a noisy ambience and on a chosen set of observed deterministic propagation parameters.

Comparison to Metric Locating

Metric or crisp locating determines a spatial or planar relationship between independently moving or residing entities (usually addressed as targets) by means of qualified methods for measuring distances. This is the topic for example of:

- Satellite positioning systems, as GPS or Galileo or,
- Real time locating systems (RTLS) as defined with ISO/IEC 19762-5 or,
- Inertial navigation systems (INS).

All of these technologies make use of a travel time measurement. Travel time measurement requires a multiplicity of measures for unambiguous locating. All of these physical methods of measuring are hampered with a challenge caused by motion and caused by transmitter population. This makes restrictions effective in time both for observation and capture and for communication of measurement data. In consequence, the pecuniary and the technical effort adjusts to physical limitations and the limited metric precision with a special aspect to operational clarity. Such balancing neglects classical terms of metric precision, prevents from over-interpreting erratic measures and provides sufficient escape.

In comparison to locating on the move, exactly determining a location with highest possible precision is the topic of geodesy and surveying. These disciplines traditionally do not deal with motion and may integrate over long time. The terms of 'locating', 'positioning' and 'navigating' or 'surveying' are commonly used in almost equivalence, hence neglecting that the sense of these terms is different concerning sensor and actor functions and motion conditions.

Radio Signal Strength Indication (RSSI) as a Metrics

For many purposes, a distinct and reliable determination of a location relative to somewhat rastered position on a floor or just a room in a building will be sufficient for sound fuzzy locating by single measurements.

Typically fuzzy locating coincides with simple power level measurement, usually configured as unilateration. A combination of multiple distance measurement, as a multilateration, based on power level measurement, appears unbalanced. The effort for multiple measures aiming at an unambiguous multilateration process will not be justified with the achievable precision of the results of power level measures.

Though leaving some ambiguity with sparse measurements, the contiguity may be assessed applying a priori knowledge. Such includes primarily tracking motion over time. As a generalized approach the fuzzy locating based on power levels measured with wireless nodes is roughly sufficient for coarse guessing a location, where an object or a person resides in contiguity to other wireless nodes in a wireless environment.

Each wireless node varies the received power level with the distance to the transmitting wireless node and this parameter can be measured. Some additional calibration parameters serve as a basis for the statistical model of the parameters of propagation in a known neighborhood of distributed wireless nodes and of other passive objects, which influence propagation. The location estimate, which approximates the reallocation of the transmitting node, will be described with the stochastic model and the monitored parameters of propagation

Physical Challenges and Restrictions

Measurement of propagation parameters is generally heavily loaded with various noise components. Such noise may be stochastic ([white noise]) or deterministic or a mixture of noises with limited bandwidth ([pink noise]).

As with all wireless systems, qualities of measurement in any one aspect contradict to qualities in some other aspects:

- Especially extension of reach collides with reducing the separation limits.
- Also increase of probing duration collides with motion speed.
- Generally the requirement for unambiguous locating requires a set of measured distances.

Such contracdictions result in challenges for systems performance and in hard restrictions concerning timely parallel availability of certain quality levels. Facing limitations as inevitable, the implementer of a locating system has to determine the operational requirements first and then has to make a choice under a set of alternatives and must scale the adjoint limitations. The outcome is always a compromise with trade offs usually in budgetary effort and in technical precision. Any chosen alternative generally will exclude certain other technical options from operational availability.

Noisy Ambience as a General Condition

In technical terms, operational environments are generally noisy. For measuring that is not a friendly ambience. Systematic and stochastic errors under operational requirements and conditions turn virtually precise measures to fuzzy metrics. The measuring results get in worse in a densely populated ambience especially in the vicinity of other electrically active objects. Even physically passive surfaces contribute to the measurement problem. All physical effects tend to contribute unsteady and non linear behavior. Hence the physical measurement errors lead to biased or erratic results under such noisy conditions.

Sequencing as a Restriction

In general, the strict sequencing of tasks appears with single tasking in one processor. Similarly the factual sequencing on only one frequency results from anti collision procedures. Both types of sequencing produce some dilation of time (with anti collision) or some dilution of location (with moving targets), while the respective wireless processes are performed by each target.

Frequency Spread as an Option

One escape from collision problems with wireless networks is the separation of the measuring and the communications processes with allocation to different frequencies, however requiring respective dual transponder capabilities. Another escape from collision problems with wireless networks is the spread of signals with stochastic coding.

Population as a Challenge

When several targets move independently in the same area and same wireless reach and also request locating independently and potentially in timely conflict using on the very same frequency for communications and for measurement, then the required measurements in one single ambience may collide. One escape is the separation of the measuring and the communications processes with allocation to different frequencies, however requiring respective dual transponder capabilities.

Motion as a Challenge

When a target moves at a certain speed, the sequential measuring of distances from such transmitter target to a set of responder targets will deliver distance data for the subsequent locations at each measuring directly back to the transmitter target. This effect is independent from architecture of the network. However, a measuring triggered from the transmitter target but performed almost in parallel by a set of receiver targets delivers a much better result under motion conditions, but requires either a server function for collecting the resulting data or requires additional response back to the triggering transceiver target. The other escape is to apply a procedure to bundle the required measurements for each target in direct sequence thus reducing one effect of motion challenge by saving the preparation times for a communications link. If not, then the competition for non-colliding transmission will lengthen the time span for each set of measurements.

Line of Sight as a Problem

In any case, line of sight is required for distance measurements. This may be eased by using auxiliary targets, but then increases the count of measurements. And the usage of auxiliary targets burdens the results with an increase of inaccuracy.

Multipath Propagation as a Problem

Multipath propagation is inevitable with wireless systems. The reception from any transmitter and the response to any transmission are both challenged by the option of multiple propagation paths. If there is none but a single cranked path, there is no desired result at all, and the option to discriminate false measures from proper measures fails completely. Typical issues with multipath propagation are fading, dither, diversity, non-linearity.

Time as a Restriction for Measuring

Even if targets do not move, time is a restriction with the performing of a locting function. The first impact is that of allowed minimal distinct time differences that define the theoretically achievable precision. This varies conceptually between phase discrimination in fractions of a cycle to be measured and full cycles to be counted. In competition for non-colliding transmission, time may appear as the main aspect with systems that use the same deterministic frequencies. Other stochastic frequency allocation may ease the thrive for results, but normally coincide with lower allowance for power.

Allowable time differences mostly vary with motion of the observed target. As for determined locating in a noisy environment several measurements are required and for disambiguation in space generally four measurements from reference points determine a target location, time appears as the sparse parameter.

Bandwidth as a Restriction

The measuring of signals with steady modulation is bound to bandwidth of the modulated carrier. The measuring of chirped signals is equivalently bound to bandwidth of the transmitted pulses. In both methods the available bandwidth limits the precision of results.

Resolution as a Restriction

Technical means for measuring offer limited resolution and respective digitizing errors. This limits again the quality of results. Any way to overcome such limits raises system cost. Hence the escape again is not in improving the technical effort, but directs to the mathematical yield.

Mathematical Requirements and Options

Mathematics serves for everything that cannot be covered by physics approaches. The assumption that a most qualified electro technical approach solves all problems arising from measurement is naïve and does not lead to sufficient results. At least thriving for best performance only at the expense of electronics is not an economized approach.

An operationally sufficient locating system will balance benefit and effort. Measurement and estimate shall take motion into account. This must not include the measuring of motion itself, but proper assessment of current and past motion to estimates. All estimate approximating the real location of the target is determined on the basis of a statistical model for the observed stochastic processes. Such model and estimation will use the set of observed propagation parameters. Some calibration data may serve as a basis for a statistical model of the propagation parameters. Such calibration is performed versus a spatial distribution of radio energy and with aspect to a known spatial distribution of corresponding targets. Other passive objects affecting propagation interfere with wireless operation and measurement.

Filtering as a Basic Requirement

Any measurement is always biased with disturbances from ambient radiation out of electrical units with switches, from other wireless units and from stationary equipment as computers. To eliminate erratic results, some estimation based on past behavior, current dynamic properties and with reference to coupling mechanisms is recommended. State of the art is e.g. extended Kalman filtering. Approaches that do not apply filtering produce no reasonable results. However, scalar filtering uses the model of residence in a fixed location or stationary motion.

Statistics as a Means for Estimation

In case of biased signals the prerequisite for filtering is some statistic estimation, which serves for eliminating the large errors and smoothes a sequence of measurement results. This may be integrated with filtering, as far as the eliminating of large errors does not bias the filtering process under any conditions.

Quadratic Equations as a Problem

The determining equations are quadratic ones, thus requiring at least one more equation (n+1) than defined by dimension (n). This leads to a minimum requirement of three equations for planar problems and four equations for spatial problems.

Over Determination as a Support

The common approach to locating calculations may be the inversion of the Euclidic distance equations. However, such deterministic approach does not serve for the balancing in over determined equation systems. The easy approach is the exploitation of Gauss' least squares principle with the multi dimensional scaling.

Wireless Coexistence

Many offered systems architectures and product offerings use license free ISM frequency bands and reside in similar channel patterns. The operating of fuzzy locating shall not compromise the communications options. Some restrictions apply not to infringe this requirement.

Technology Approaches and Options

The second step after scalar calculation is the involvement of model data according to the dimensionality of the motion. If reference is made to targets in other planes but the plane on which the moving targets may operate, such model must be a three dimensional model. For model based operations, there are several options.

Coincident Locating as the Initial Option

Imagine a worker operating with a handheld reader of any type. The person is skilled to capture the identity of an object and report the capturing with time stamp. Such report discloses the location of capture as far as this information is reported in contents. The mandatory condition could be some automatic means to capture the location at the moment of identifying. In all other cases the quality and reliability of the location report limits the validity of the e.g. vocally reported data.

In all implementations of automatic data acquisition and locating systems the option of locating a handheld reader in the moment of manual triggering shall be foreseen as the fall back option. Otherwise the robustness of automatic data acquisition systems operation is bound to availability of automatic operation only.

Choke Point Locating as the Poorest Option

A choke point is a static bottleneck in process flow designs. There the passage of individuals and/or objects may indicate the identity of such entity to a steadily installed identifier unit. This approach under all conditions is restricted to just one location. Politics and Sales force may describe that as locating, but it is definitively still just identifying.

Power Mapping as a First Poor Option

Propagation of radio signals happens according to Maxwell's equations and includes attenuation in atmosphere proportional to distance, Such concept is the basis for power mapping. The irregularities from local ambient conditions may be taken into account by power measurement in the operational area to correct the theoretically linear attenuation with distance. However, this approach does not work with an accuracy of better than 10% of the calculated distances in the range of propagation, thus leading to accuracies in the range of some meters.

Time Distance Equivalence for Radiation

Propagation of radio signals happens according to Maxwell's equations and includes travel time in atmosphere proportional to distance, Such concept is the basis for precise distance measurement. The irregularities from local ambient conditions are not dominant, thus the approach is more precise than power measurement. So this approach works with an accuracy of better than 1% of the calculated distances in the range of propagation, thus leading to accuracies in the range of some centimeters. However, this approach serves as well for line of sight propagation as for indirect reflected propagation.

Space Model as a Strong Option

To escape the biasing with secondary paths, there must be some reasoning that excludes the physically impossible locations from sets of results from locating. Simply, all calculated locations in material will be assessed as erratic, all calculated locations at distances not possible with inherent speed limits will be assessed as erratic and all locations above ground will be assessed impossible for floor operation. The requirement for space modeling leads to depicting the operational planes different from the limits to such planes, as walls, racks, and other installations.

Statistical Model to Exploit the Measurements

There is no chance to base stable results for location on single measurements. Statistics allow for:

- Combining subsequent measurement results to form a track.
- Smoothing a single location from subsequent measurement results.
- Iterating stable results from coarse first estimates.

The methods for computing a set of results are described in context of various applications not just with locating technologies.

Fuzzy Reasoning with Discrete Spatial Compartments

As far as locating just has to support discrimination of rooms where a target may reside, the continuous model approaches may be combined with reasoning procedures to eliminate improbable results and to exclude operationally invalid locations from potential depicting a scenery. The known methods of inference apply to such processing.

Geometric Mapping Contributions to Reasoning

As far as the ambient operational conditions are stable, a geometric mapping of the neighborhood may support the reasoning. Then all massive obstacles describe the residual space of operation. As well such mapping will support the systematic and well determined consideration of multipath propagation effects. Hence geometric mapping derives the major gain compared to Bayes' estimators.

Adaptive Approaches

A common approach as preparatory mapping requires steady conditions and a constant ambience. This crucial condition is not fulfilled in dynamic operational theatres. However, a robust solution will always detect and investigate the actual conditions and reconnoiter the present ambience. For robot navigation, the methods of adaptive systems design, hence application of learning functions, is state of the art. However, adaptation requires time. A fully adaptive solution not applying a priori knowledge will be rather slow and will show limited dynamics. A balanced combination of adaptive functions will allow for best performance in a generally known ambience and cope for all changes that occurred after last encounter.

Operational Requirements

Locating arises from operational challenges. Traditional understanding of well kept enterprises with well educated staff is undermined with a thriving for reduced skills to achieve lesser cost. In result and in addition with continuing socio-economical disparities the processes and objects under control are threatened by negligence, fraud and theft.

Evidence

A simple indication of presence is given with the signal used for locating an object or a person. However, as presence may be temporary, a time stamp is required to adduce evidence in retrospective.

Cooperation Requirement

Persons carrying transponders or tagged objects with transponders might not be willing to be observed though having agreed earlier to this process. Then cooperation may be technically required, but individually denied. The robustness of the detection hence shall not be dependent to such cooperation. Especially covering the transponder or tearing off the transponder or otherwise tampering must be sensed automatically.

Proof of Presence

The presence of an object or a person in an operational vicinity is a strong demand.

Absence of required resources generally affects planned processes. Therefore the proof of presence may be performed as far as possible before binding of additional resources happens.

Co-locating of Staff

Specially team work is bound to availability of required staff. The persons involved in a scheduled operation are well skilled to determine who is missing, but locating the missing parties is not that easy and may be strongly improved by system support.

Discrimination of Rooms

To allow for operation the respective room shall coincide with the scheduled action. Any request to operate under restrictions outside the planned confined area is suspect and may challenge security of processes and of secured knowledge. Locating the acting entities in the named confinement contributes to fulfilling security requirements.

Coincidence of Presence and Challenge

A person may try to access data, material or other resources outside the well secured rooms or areas. However, control may not always secure the subordination to given orders. Locating simply may confirm the request is a basic feature.

Quality Requirements

The listed terms will show that the definition of desired precision and accuracy, of repeatability and delays alone does not comply with a proper definition of requirements under aspects of cost. However, other terms of quality apply without restrictions.

Tamper Proof Identity

Basic requirement for any means to support locating is a tamper proof inherent identity of the carrying target with secured access. The secrecy of the identity prevents from plump copying threat and the tamper protection prevents from manipulating the target.

Self Identifying Authenticity

Persons who pursue to access data and applications normally authenticate themselves. Such authentication is generally bound to known locations, where the persons are authorized to perform work. Locating persons when they challenge authorization procedures is and advantage to prevent from fraud and theft.

Object Identifying Security

Numerous means are known to identify objects. Normally the location where hand held units are operated are just roughly determined by the access point where connection to network is made. However such locating is still an improvement in many operations to secure knowledge about whereabouts of objects upon identifying.

Tracking Capability

In larger context of spatially distributed services and especially in logistics, numerous objects are in use in parallel, in different locations or on the move. Especially with transportation whereabouts of objects are understood as an essential to achieve high quality of service. As far as trust is with the forwarder, no problem exists on the journey and locating may happen just on leave and upon arrival. But third party infringements may collide with this assumption and generate a demand for permanent tracking on the journey as well. Then fuzzy locating is an economized and sufficient approach, which shall not provide location data with high metric accuracies, but status information with checkable and justifiable evidence.

Tracing Capability

In case single objects are lost, the capability to trace the whereabouts is another option to get access to the missing target again. However, this tracing is performed on yet available data and no means will deliver the data from the past without respective precautions. Especially in transportation whereabouts of lost objects are understood as an essential to retrieve the missing belongings.

Alert on Deviation

Easily any deviation from planned course, set route and scheduled arrival may lead to an alert. This requires timely locating and comparison of captured data with planning.

FUZZY COGNITIVE MAP

A fuzzy cognitive map (FCM) is a cognitive map within which the relations between the elements (e.g. concepts, events, project resources) of a "mental landscape" can be used to compute the "strength of impact" of these elements. Fuzzy cognitive maps were introduced by Bart Kosko. Ron Axelrod introduced cognitive maps as a formal way of representing social scientific knowledge and modeling decision making in social and political systems, then brought in the computation fuzzy logic.

Fuzzy cognitive maps are signed fuzzy digraphs. They may look at first blush like Hasse diagrams but they are not. Spreadsheets or tables are used to map FCMs into matrices for further computation. FCM is a technique used for causal knowledge acquisition and representation, it supports causal knowledge reasoning process and belong to the neuro-fuzzy system that aim at solving decision making problems, modeling and simulate complex systems. Learning algorithms have been proposed for training and updating FCMs weights mostly based on ideas coming from the field of Artificial Neural Networks. Adaptation and learning methodologies used to adapt the FCM model and adjust its weights. Kosko and Dickerson suggested the Differential Hebbian Learning (DHL) to train FCM. There have been proposed algorithms based on the initial Hebbian algorithm; others algorithms come from the field of genetic algorithms, swarm intelligence and evolutionary computation. Learning algorithms are used to overcome the shortcomings that the traditional FCM present i.e. decreasing the human intervention by suggested automated FCM candidates; or by activating only the most relevant concepts every execution time; or by making models more transparent and dynamic.

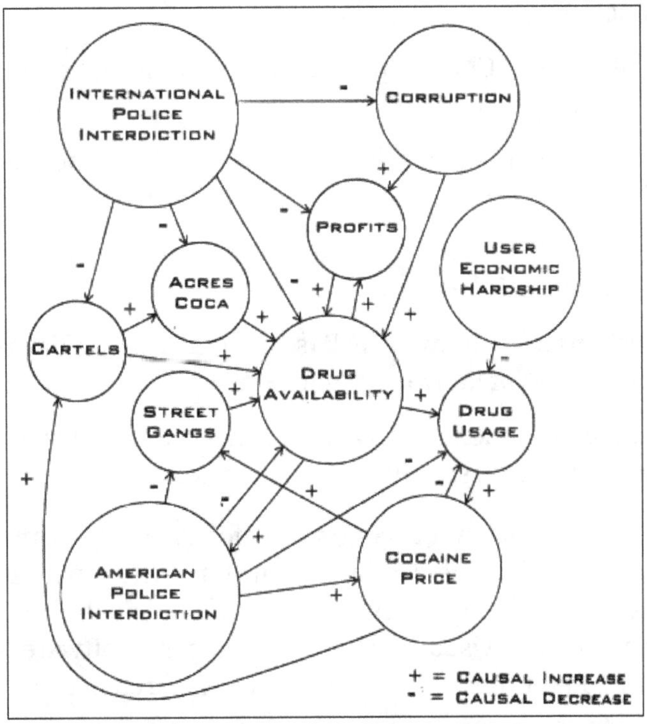

Rod Tabers FCM depicting eleven factors of the drug market.

Fuzzy cognitive maps (FCMs) have gained considerable research interest due to their ability in representing structured knowledge and model complex systems in various fields. This growing interest led to the need for enhancement and making more reliable models that can better represent real situations. A first simple application of FCMs is described by William R. Taylor, by analyzing the war in Afghanistan and Iraq. As an example, one FCM quoted from Rod Taber describes 11 factors of the American cocaine market and the relations between these factors. For computations, Taylor uses pentavalent logic (scalar values out of {-1,-0.5,0,+0.5,+1}). That particular map of Taber uses trivalent logic (scalar values out of {-1,0,+1}).

While applications in social sciences introduced FCMs to the public, they are used in a much wider range of applications, which all have to deal with creating and using models of uncertainty and complex processes and systems. Examples:

- In business FCMs can be used for product planning and decision support.
- In economics, FCMs support the use of game theory in more complex settings.
- In education for modeling Critical Success Factors of Learning Management Systems.
- In medical applications to model systems, provide diagnosis, develop decision support systems and medical assessment.
- In engineering for modeling and control mainly of complex systems and reliability engineering.
- In project planning FCMs help to analyze the mutual dependencies between project resources.
- In robotics FCMs support machines to develop fuzzy models of their environments and to use these models to make crisp decisions.
- In computer assisted learning FCMs enable computers to check whether students understand their lessons.
- In expert systems a few or many FCMs can be aggregated into one FCM in order to process estimates of knowledgeable persons.
- In IT project management, a FCM-based methodology helps to success modelling, risk analysis and assessment, IT scenarios.

FCMappers is an international online community for the analysis and the visualization of fuzzy cognitive maps. FCMappers offer support for starting with FCM and also provide a Microsoft Excel-based tool that is able to check and analyse FCMs. The output is saved as Pajek file and can be visualized within third party software like Pajek, Visone, etc. They also offer to adapt the software to specific research needs.

Additional FCM software tools, such as Mental Modeler, have recently been developed as a decision-support tool for use in social science research, collaborative decision-making, and natural resource planning.

FUZZY CONTROL

Fuzzy control is a methodology to represent and implement a (smart) human's knowledge about how to control a system. A fuzzy controller is shown in figure. The fuzzy controller has several components:

- The rule-base is a set of rules about how to control.

- Fuzzification is the process of transforming the numeric inputs into a form that can be used by the inference mechanism.
- The inference mechanism uses information about the current inputs (formed by fuzzification), decides which rules apply in the current situation, and forms conclusions about what the plant input should be.
- Defuzzification converts the conclusions reached by the inference mechanism into a numeric input for the plant.

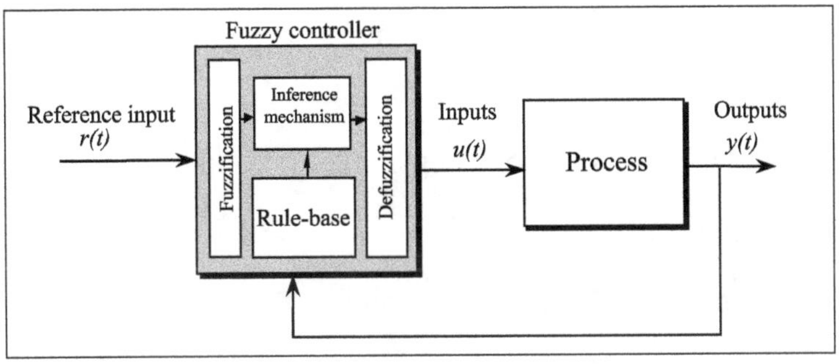

Fuzzy control system.

Fuzzy Control Design

As an example, consider the tanker ship steering application in figure below, where the ship is traveling in the x direction at a heading ψ and is steered by the rudder input δ. Here, we seek to develop the control system in figure below by specifying a fuzzy controller that would emulate how a ship captain would steer the ship. Here, if ψ_r is the desired heading, $e = \psi_r - \psi$ and $c = \dot{e}$.

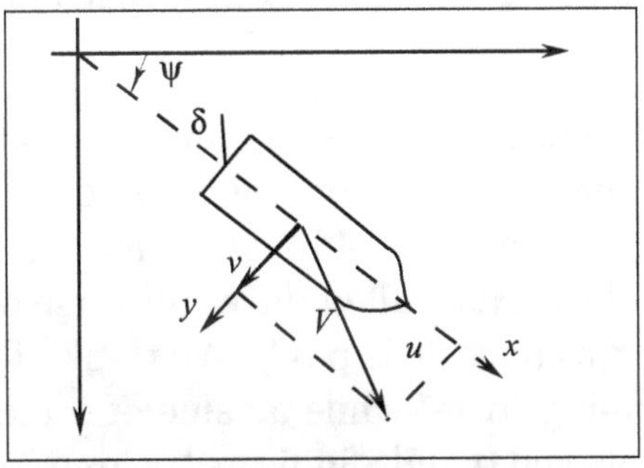

Tanker ship steering problem.

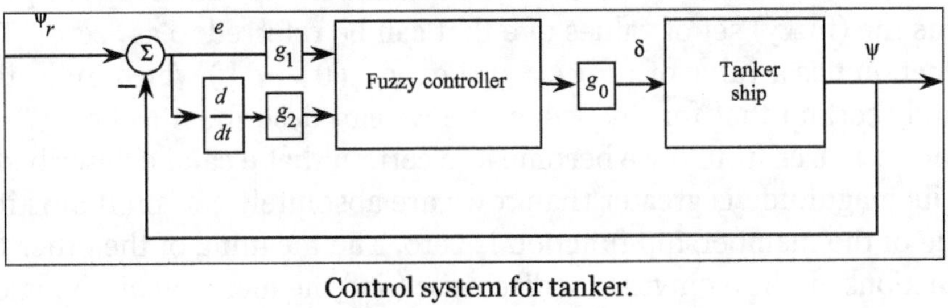

Control system for tanker.

The design of the fuzzy controller essentially amounts to choosing a set of rules ("rule base"), where each rule represents knowledge that the captain has about how to steer. Consider the following set of rules:

- If e is neg and c is neg Then δ is poslarge,
- If e is neg and c is zero Then δ is possmall,
- If e is neg and c is pos Then δ is zero,
- If e is zero and c is neg Then δ is possmall,
- If e is zero and c is zero Then δ is zero,
- If e is zero and c is pos Then δ is negsmall,
- If e is pos and c is neg Then δ is zero,
- If e is pos and c is zero Then δ is negsmall,
- If e is pos and c is pos Then δ is neglarge.

Here, "neg" means negative, "poslarge" means positive and large, and the others have analogous meanings. What do these rules mean? Rule 5 says that the heading is good so let the rudder input be zero. For Rule 1:

- "e is neg" means that ψ is greater than ψ_r.
- "c is neg" means that ψ is moving away from ψ_r (if ψ_r is fixed).
- In this case we need a large positive rudder angle to get the ship heading in the direction of ψ_r.

The other rules can be explained in a similar fashion. What, precisely, do we (or the captain) mean by, for example, "e is pos," or "c is zero," or "δ is poslarge"? We quantify the meanings with "fuzzy sets" ("membership functions"), as shown in figure. Here, the membership functions on the e axis (called the e "universe of discourse") quantify the meanings of the various terms (e.g., "e is pos"). We think of the membership function having a value of 1 as meaning "true" while a value of 0 means "false." Values of the membership function in between 0 and 1 indicate "degrees of certainty." For instance, for the e universe of discourse the triangular membership function that peaks at e = 0 represents the (fuzzy) set of values of e that can be referred to as "zero." This membership function has a value of 1 for e = 0 (i.e., $\mu_{zero}(0) = 1$) which indicates that we are absolutely certain that for this value of e we can describe it as being "zero." As e increases or decreases from 0 we become less certain that e can be described as "zero" and when its magnitude is greater than π we are absolutely certain that is it not zero, so the value of the membership function is zero. The meaning of the other two membership functions on the e universe of discourse (and the membership functions on the changein-error universe of discourse) can be described in a similar way. The membership functions on the δ universe of discourse are called "singletons." They represent the case where we are only certain that a value of δ is, for example, "possmall" if it takes on only one value, in this case $\frac{40\pi}{180}$, and for any other value of δ we are certain that it is not "possmall." Finally, notice that figure below shows the relationship between the scaling gains in figure and the scaling of the universes of discourse (notice that for the inputs there is an inverse relationship since an increase an input scaling gain corresponds to making, for instance, the meaning of "zero" correspond to smaller values).

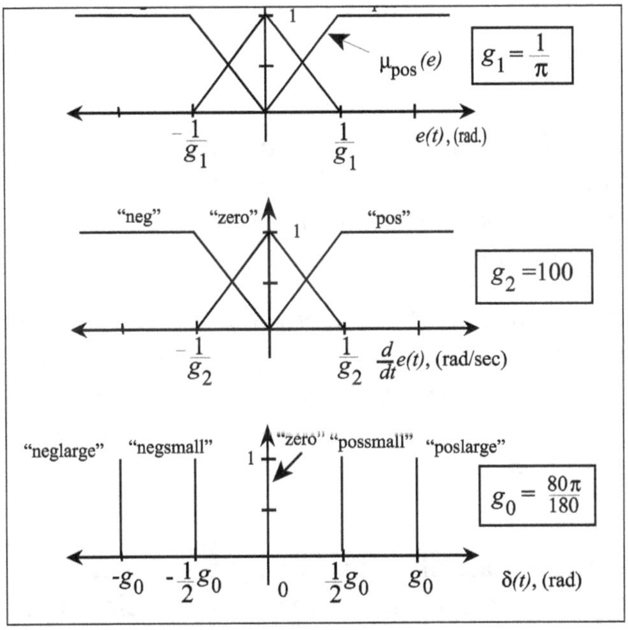

Membership functions for inputs and output.

It is important to emphasize that other membership function types (shapes) are possible; it is up to the designer to pick ones that accurately represent the best ideas about how to control the plant. "Fuzzification" is simply the act of finding, e.g., $\mu_{pos}(e)$ for a specific value of e.

First, we use "fuzzy logic" to quantify the conjunctions in the premises of the rules. For instance, the premise of Rule 2 is, "*e* is neg and c is zero."

Let $\mu_{neg}(e)$ and $\mu_{zero}(c)$ denote the respective membership functions of each of the two terms in the premise of Rule 2. Then, the premise certainty for Rule 2 can be defined by:

$$\mu_{premise(2)} = \min\{\mu_{neg}(e),\ \mu_{zero}(c)\}$$

Why? Think about the conjunction of two uncertain statements. The certainty of the assertion of two things is the certainty of the least certain statement.

In general, more than one $\mu_{premise(i)}$ will be nonzero at a time so more than one rule is "on" (applicable) at every time. Each rule that is "on" can contribute to making a recommendation about how to control the plant and generally ones that are more on (i.e., have $\mu_{premise(i)}$ closer to one) should contribute more to the conclusion. This completes the description of the inference mechanism.

Defuzzification involves combining the conclusions of all the rules. "Center-average" defuzzification uses,

$$\delta = \frac{\sum_{i=1}^{9} b_i \mu_{\text{premise(i)}}}{\sum_{i=1}^{9} \mu_{\text{premise(i)}}}$$

Here b_i is the position of the center of the output membership function for the ith rule (i.e., the position of the singleton). This is simply a weighted average of the conclusions. This completes the description of a simple fuzzy controller (and notice that we did not use a mathematical model in its construction).

There are many extensions to the fuzzy controller that we describe above. There are other ways to quantify the "and" with fuzzy logic, other inference approaches, other defuzzification methods, "Takagi-Sugeno" fuzzy systems, and multi-input multi-output fuzzy systems.

Ship Example

Using a nonlinear model for a tanker ship we get the response in figure (tuned using ideas from how you tune a proportional-derivative controller; notice that the values of $g_1 = 2/\pi$, $g_2 = 250$, and $g_0 = 8\pi/18$ are different than the first guess values and the controller surface. The control surface shows that there is nothing mystical about the fuzzy controller! It is simply a static (i.e., memoryless) nonlinear map. For real-world applications most often the surface will have been shaped by the rules to have interesting nonlinearities.

Design Concerns

There are several design concerns that one encounters when constructing a fuzzy controller. First, it is generally important to have a very good understanding of the control problem, including the plant dynamics and closed-loop specifications. Second, it is important to construct the rulebase very carefully. If you do not tell the controller how to properly control the plant, it cannot succeed! Third, for practical applications you can run into problems with controller complexity since the number of rules used grows exponentially with the number of inputs to the controller, if you use all possible combinations of rules (however, note that the number of rules on at any one time grows much slower for the ship example). As with conventional controllers there are always concerns about the effects of disturbances and noise on, for example, tracking error (just because it is a fuzzy controller does not mean that it is automatically a "robust" controller). Indeed, analysis of robustness properties, along with stablity, steady state tracking error, and limit cycles can be quite important for some applications. Since the fuzzy controller is a nonlinear controller, the current methods in nonlinear analysis apply to fuzzy control systems.

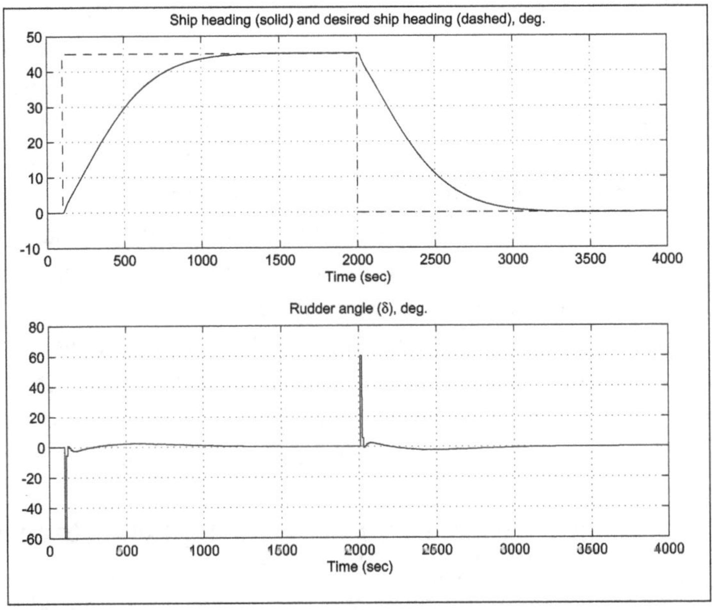

Response of fuzzy control system for tanker heading regulation

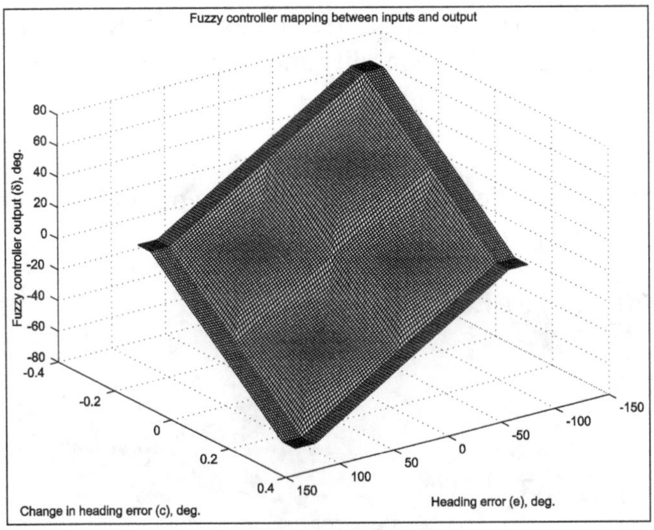

Fuzzy controller surface.

Fuzzy Control using Fuzzy Estimators and Feedback Linearization

In recent years the area of control for nonlinear systems has been the subject of many studies. Computational advances have enabled more complex applications to provide solutions to nonlinear problems. The method is applicable to a class of nonlinear system governed by state equations of the form $\dot{x} = f(x) + g(x)u$, where $f(x)$ and $g(x)$ represent the nonlinearities of the states. Some classic control applications can be described by this specific class of nonlinear systems, for example, inverted pendulum, conic vessel, Continuously Stirred Tank Reactor (CSTR), and magnetic levitation system. The direct application of linear control techniques (e.g.: PID) may not be efficient for this class of systems. On the other hand, classical linearization methods may lead an adequate performance when the model is accurate, even though normally limited around the point in which the linearization took place. However, when model uncertainties are considerable the implementation of the control law becomes difficult or unpractical. In order to deal with model uncertainties an adaptive controller is used. The controller implements two basic ideas. First, the technique of exact feedback linearization is used to handle the nonlinearities of the system. Second, the control law formulation in presence of model uncertainties is made with the estimates of the nonlinear functions $f(x)$ and $g(x)$. The adaptation mechanism is used to adjust the vector of parameters θ_f and θ_g in a singleton fuzzyfier zero-order Takagi-Sugeno-Kang (TSK) structure that provides estimates in the form $f(x \mid \hat{\theta}_f)$ and $g(x \mid \hat{\theta}_g)$. The fuzzy logic system is built with a product-inference rule, center average defuzzyfier, and Gaussian membership functions. One of the most important contributions of the adaptive scheme used here is the real convergence of the estimates $f(x \mid \hat{\theta}_f)$ and $g(x \mid \hat{\theta}_g)$ to their optimal values $f(x \mid \theta_f^*)$ and $f(x \mid \theta_g^*)$ while keeping the tracking error with respect to a reference signal within a compact set. Convergence properties are investigated using Lyapunov candidate functions. In order to illustrate the methodology, a nonlinear and open-loop unstable magnetic levitation system is used as an example. Experimental tests in a real plant were conducted to check the reliability and robustness of the proposed algorithm.

A Specific Structure

The proposed method is applicable to a specific class of nonlinear system that can be described by state equations of form:

$$\begin{cases} \dot{x} = Ax + B\left[f(x) + g(x)u\right] \\ y = Cx \end{cases}$$

where,

$$A = \begin{bmatrix} 0 & 1 & 0 & \cdots & 0 \\ 0 & 0 & 1 & \cdots & 0 \\ \vdots & \vdots & \vdots & \ddots & \vdots \\ 0 & 0 & 0 & 0 & 1 \\ 0 & 0 & 0 & 0 & 0 \end{bmatrix}_{n \times n}$$

$$B = \begin{bmatrix} 0 & \cdots & 0 & 1 \end{bmatrix}_{1\times n}^{T}$$
$$C = \begin{bmatrix} 1 & 0 & \cdots & 0 \end{bmatrix}_{1\times n}$$

n is the dimension of the system.

$x \in X \subset \Re^n$, are the states of the system $f(x): \Re^n \to \Re s.t. f(x)$ is continuous and $f(x) \in U_f \subset \Re, \forall x \in X, |f(x)| \le f^U, \forall x \in X$.

$g(x): \Re^n \to \Re s.t.\ g(x)$ is continuous and $g(x) \in U_g \subset \Re, \forall x \in X$ $0 < \zeta < |g^L| \le |g(x)| \le |g^U|$, $\forall x \in X$, for a constant $\zeta > 0$ u is the control signal y is the output signal or, in an equivalent form:

$$\begin{cases} \dot{x}_1 = x_2 \\ \dot{x}_2 = x_3 \\ \vdots \\ \dot{x}_n = f(x) + g(x)u \end{cases}$$
$$y = Cx = x_1$$

where, X, U_f and U_g are compact sets, sign($g(x)$) = sign(g^L) = sign(g^U) and x = 0 is an interior point of X.

Examples of Nonlinear Systems

Once the class of nonlinear systems is defined, it is important to select a plant to highlight the controller properties. As this structure is very common in control applications, some classic systems can be written using equation $\begin{cases} \dot{x} = Ax + B\left[f(x) + g(x)u\right] \\ y = Cx \end{cases}$ format:

Control Law

Consider a nonlinear system described by state equations in the form of equation $\begin{cases} \dot{x} = Ax + B\left[f(x) + g(x)u\right] \\ y = Cx \end{cases}$. Let the control objective be to track a reference signal $r(t) \in S \subset \Re$, where S is a compact set of possible references which can be supplied to the nonlinear plant.

Exact feedback linearization technique can be used here for exacting cancellation of the nonlinear functions $f(x)$ and $g(x)$. If one assumes that the functions $f(x)$ and $g(x)$ are known, then an interesting structure for the control law would be,

$$u = \frac{1}{g(x)}\left[-f(x) + D(x)\right]$$

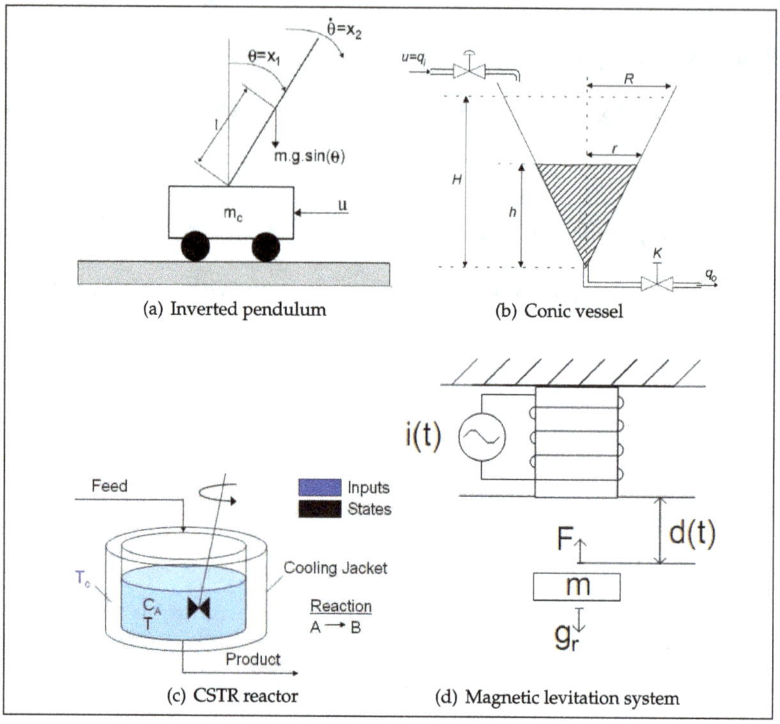

Examples of nonlinear systems.

Where, D is to be selected in the design phase.

By substituting above equation in the nonlinear system equation:
$\begin{cases} \dot{x} = Ax + B\left[f(x) + g(x)u\right] \\ y = Cx \end{cases}$ one gets, $\dot{x} = Ax + BD(x)$ so that $\dot{x}_n = D(x)$, which can be chosen as the desired dynamics for the controlled system.

Example: Let a gain vector K be $K = [K_1, \ldots, K_n] \in R^n$ and consider $D(x) = r - K(x)$. The proposed control law becomes:

$$u = \frac{1}{g(x)}\left[-f(x) + r - Kx\right]$$

By substituting above equation in the nonlinear system in equation $\begin{cases} \dot{x} = Ax + B\left[f(x) + g(x)u\right] \\ y = Cx \end{cases}$, one gets $\dot{x} = (A - BK)x + B\,r$.

Which yields a linear dynamic. In this case, the stability can be verified directly by computing the eingevalues of the matrix (A – BK).

Based on equation $u = \frac{1}{g(x)}\left[1 - f(x) + D(x)\right]$, a large number of control laws can be suggested. For instance, the control laws used in and, as well as the control law proposed in equation $u = \frac{1}{g(x)}\left[-f(x) + r - Kx\right]$ are particular cases of equation $u = \frac{1}{g(x)}\left[1 - f(x) + D(x)\right]$.

Without loss of generality, r = 0 is considered for stability analysis and the nominal system is considered to be of the form,

$$\dot{x} = (A - BK)x$$

Mathematical Requirements for Fuzzy Logic Control Application

Fuzzy Structure

Since the fuzzy logic was proposed by L.A. Zadeh, a lot of different fuzzy inference engines have been suggested. Other researchers have also been used with success in a variety of applications for control of nonlinear systems. Based on previous researches, which propose the Takagi-Sugeno-Kang fuzzy structure (TSK), this work makes use of a fuzzy logic system with product-inference rule, center average defuzzyfier, and Gaussian membership functions.

Consider singleton fuzzifier and let a fuzzy system be composed by R rules, each one of them of the form

IF x_1 is A_1^j and x_n is A_n^j THEN y is B_j

where, $x = [x_1, \ldots, x_n] \in \Re^n$ is the input vector, $\{A_1^j, \ldots . A_n^j\} / B_j$ are the input/output fuzzy sets related to the jth rule (j = 1, . . . , R), and y is the fuzzy output.

Consider y_j as the point in which B_j is maximum $\left(\mu_{Bj}(y_j) = 1\right)$ and define θ as the vector of parameters of the form,

$$\theta^T = [y_1, \ldots, y_R]$$

Therefore, the fuzzy output can be expressed as:

$$y = \theta^T . W(x)$$

Where,

$$W(x) = [W_1(x), \ldots, W_R(x)]^T$$

$$W_j(x) = \frac{\Pi_{k=1}^{n} \mu_{A_k^j}(x_k)}{\Sigma_{j=1}^{R}\left(\Pi_{k=1}^{n} \mu_{A_k^j}(x_k)\right)}, j = 1.....R$$

and,

$$W_j(x) \in [0, 1]$$

is usually called the weight of the jth rule. The scheme is shown in figure below.

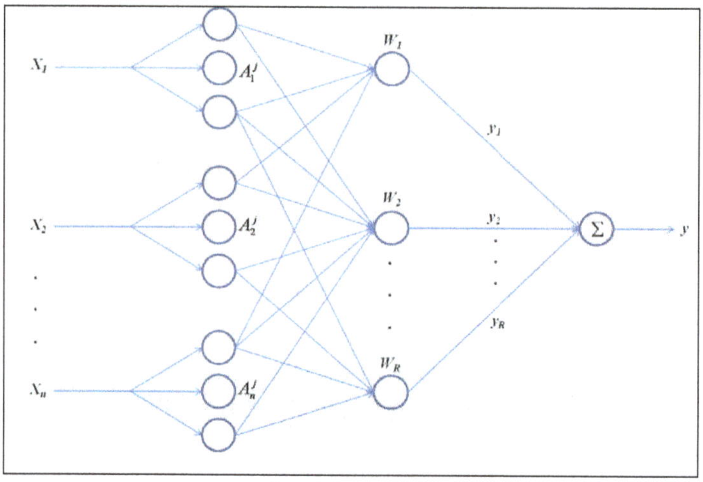

Zero-order TSK fuzzy structure.

Fuzzy as Universal Approximators

For any given real continuous function $f(x)$ on a compact set $x \in X \subset \Re^n$ and arbitrary $\varepsilon > 0$, there exists a fuzzy logic system $f(x \mid \theta^*)$ in the form of $y = \theta^T.W(x)$ such that,

$$\max_{x \in X} \left| f(x) - f\left(x \mid \theta^*\right) \right| < \varepsilon, \forall x$$

The fuzzy estimates are then used in the control law which may, for instance, be based on the exact feedback linearization techniques. However, it is shown by an example that the convergence of the estimates may not be achieved, even though the system exhibits good reference tracking properties.

Notice that the control law in equation $u = \frac{1}{g(x)}\left[-f(x) + r - Kx\right]$ can not be implemented because the functions $f(x)$ and $g(x)$ are unknown and must be estimated. The idea is to construct fuzzy structures to generate the estimates $f\left(x \mid \hat{\theta}_f\right)$ and $f\left(x \mid \hat{\theta}_g\right)$, where $\hat{\theta}_f$ and $\hat{\theta}_g$ are the respective vector of parameters.

Here, $\begin{cases} \dot{x} = Ax + B\left[f(x) + g(x)u\right] \\ y = Cx \end{cases}$ uses the fuzzy structures of form,

$$f\left(x \mid \hat{\theta}_f\right) = \hat{\theta}_f^T W(x)$$

$$f\left(x \mid \hat{\theta}_g\right) = \hat{\theta}_g^T W(x)$$

where W(x) is associated with the antecedent part of the rules. Moreover, the parameters $\hat{\theta}_f$ and $\hat{\theta}_f$ are obtained using an adaptation scheme. Thus, the implementable version of the control law in $u = \frac{1}{g(x)}\left[-f(x) + r - Kx\right]$ becomes,

$$u = \frac{1}{g\left(x \mid \hat{\theta}g\right)}\left[-f\left(x \mid \hat{\theta}f\right) + r - Kx\right]$$

FUZZY CONTROL SYSTEM

A fuzzy control system is a control system based on fuzzy logic—a mathematical system that analyzes analog input values in terms of logical variables that take on continuous values between 0 and 1, in contrast to classical or digital logic, which operates on discrete values of either 1 or 0 (true or false, respectively).

Fuzzy logic is widely used in machine control. The term "fuzzy" refers to the fact that the logic involved can deal with concepts that cannot be expressed as the "true" or "false" but rather as "partially true". Although alternative approaches such as genetic algorithms and neural networks can perform just as well as fuzzy logic in many cases, fuzzy logic has the advantage that the solution to the problem can be cast in terms that human operators can understand, so that their experience can be used in the design of the controller. This makes it easier to mechanize tasks that are already successfully performed by humans.

Fuzzy logic was first proposed by Lotfi A. Zadeh of the University of California at Berkeley in a 1965 paper. He elaborated on his ideas in a 1973 paper that introduced the concept of "linguistic variables", which in this topic equates to a variable defined as a fuzzy set. Other research followed, with the first industrial application, a cement kiln built in Denmark, coming on line in 1975.

Fuzzy systems were initially implemented in Japan:

- Interest in fuzzy systems was sparked by Seiji Yasunobu and Soji Miyamoto of Hitachi, who in 1985 provided simulations that demonstrated the feasibility of fuzzy control systems for the Sendai Subway. Their ideas were adopted, and fuzzy systems were used to control accelerating, braking, and stopping when the Namboku Line opened in 1987.

- In 1987, Takeshi Yamakawa demonstrated the use of fuzzy control, through a set of simple dedicated fuzzy logic chips, in an "inverted pendulum" experiment. This is a classic control problem, in which a vehicle tries to keep a pole mounted on its top by a hinge upright by moving back and forth. Yamakawa subsequently made the demonstration more sophisticated by mounting a wine glass containing water and even a live mouse to the top of the pendulum: the system maintained stability in both cases. Yamakawa eventually went on to organize his own fuzzy-systems research lab to help exploit his patents in the field.

- Japanese engineers subsequently developed a wide range of fuzzy systems for both industrial and consumer applications. In 1988 Japan established the Laboratory for International Fuzzy Engineering (LIFE), a cooperative arrangement between 48 companies to pursue fuzzy research. The automotive company Volkswagen was the only foreign corporate member of LIFE, dispatching a researcher for a duration of three years.

- Japanese consumer goods often incorporate fuzzy systems. Matsushita vacuum cleaners use microcontrollers running fuzzy algorithms to interrogate dust sensors and adjust suction power accordingly. Hitachi washing machines use fuzzy controllers to load-weight, fabric-mix, and dirt sensors and automatically set the wash cycle for the best use of power, water, and detergent.

- Canon developed an autofocusing camera that uses a charge-coupled device (CCD) to measure the clarity of the image in six regions of its field of view and use the information provided to determine if the image is in focus. It also tracks the rate of change of lens movement during focusing, and controls its speed to prevent overshoot. The camera's fuzzy control system uses 12 inputs: 6 to obtain the current clarity data provided by the CCD and 6 to measure the rate of change of lens movement. The output is the position of the lens. The fuzzy control system uses 13 rules and requires 1.1 kilobytes of memory.

- An industrial air conditioner designed by Mitsubishi uses 25 heating rules and 25 cooling rules. A temperature sensor provides input, with control outputs fed to an inverter, a compressor valve, and a fan motor. Compared to the previous design, the fuzzy controller heats and cools five times faster, reduces power consumption by 24%, increases temperature stability by a factor of two, and uses fewer sensors.

- Other applications investigated or implemented include: character and handwriting recognition; optical fuzzy systems; robots, including one for making

Japanese flower arrangements; voice-controlled robot helicopters (hovering is a "balancing act" rather similar to the inverted pendulum problem); rehabilitation robotics to provide patient-specific solutions (e.g. to control heart rate and blood pressure); control of flow of powders in film manufacture; elevator systems; and so on.

Work on fuzzy systems is also proceeding in the United State and Europe, although on a less extensive scale than in Japan:

- The US Environmental Protection Agency has investigated fuzzy control for energy-efficient motors, and NASA has studied fuzzy control for automated space docking: simulations show that a fuzzy control system can greatly reduce fuel consumption.
- Firms such as Boeing, General Motors, Allen-Bradley, Chrysler, Eaton, and Whirlpool have worked on fuzzy logic for use in low-power refrigerators, improved automotive transmissions, and energy-efficient electric motors.
- In 1995 Maytag introduced an "intelligent" dishwasher based on a fuzzy controller and a "one-stop sensing module" that combines a thermistor, for temperature measurement; a conductivity sensor, to measure detergent level from the ions present in the wash; a turbidity sensor that measures scattered and transmitted light to measure the soiling of the wash; and a magnetostrictive sensor to read spin rate. The system determines the optimum wash cycle for any load to obtain the best results with the least amount of energy, detergent, and water. It even adjusts for dried-on foods by tracking the last time the door was opened, and estimates the number of dishes by the number of times the door was opened.

Research and development is also continuing on fuzzy applications in software, as opposed to firmware, design, including fuzzy expert systems and integration of fuzzy logic with neural-network and so-called adaptive "genetic" software systems, with the ultimate goal of building "self-learning" fuzzy-control systems. These systems can be employed to control complex, nonlinear dynamic plants, for example, human body.

Fuzzy Sets

The input variables in a fuzzy control system are in general mapped by sets of membership functions similar to this, known as "fuzzy sets". The process of converting a crisp input value to a fuzzy value is called "fuzzification".

A control system may also have various types of switch, or "ON-OFF", inputs along with its analog inputs, and such switch inputs of course will always have a truth value equal to either 1 or 0, but the scheme can deal with them as simplified fuzzy functions that happen to be either one value or another.

Given "mappings" of input variables into membership functions and truth values, the microcontroller then makes decisions for what action to take, based on a set of "rules", each of the form:

```
IF brake temperature IS warm AND speed IS not very fast
THEN brake pressure IS slightly decreased.
```

In this example, the two input variables are "brake temperature" and "speed" that have values defined as fuzzy sets. The output variable, "brake pressure" is also defined by a fuzzy set that can have values like "static" or "slightly increased" or "slightly decreased" etc.

Fuzzy Control

Fuzzy controllers are very simple conceptually. They consist of an input stage, a processing stage, and an output stage. The input stage maps sensor or other inputs, such as switches, thumbwheels, and so on, to the appropriate membership functions and truth values. The processing stage invokes each appropriate rule and generates a result for each, then combines the results of the rules. Finally, the output stage converts the combined result back into a specific control output value.

The most common shape of membership functions is triangular, although trapezoidal and bell curves are also used, but the shape is generally less important than the number of curves and their placement. From three to seven curves are generally appropriate to cover the required range of an input value, or the "universe of discourse" in fuzzy jargon.

The processing stage is based on a collection of logic rules in the form of IF-THEN statements, where the IF part is called the "antecedent" and the THEN part is called the "consequent". Typical fuzzy control systems have dozens of rules.

Consider a rule for a thermostat:

```
IF (temperature is "cold") THEN turn (heater is "high")
```

This rule uses the truth value of the "temperature" input, which is some truth value of "cold", to generate a result in the fuzzy set for the "heater" output, which is some value of "high". This result is used with the results of other rules to finally generate the crisp composite output. Obviously, the greater the truth value of "cold", the higher the truth value of "high", though this does not necessarily mean that the output itself will be set to "high" since this is only one rule among many. In some cases, the membership functions can be modified by "hedges" that are equivalent to adverbs. Common hedges include "about", "near", "close to", "approximately", "very", "slightly", "too", "extremely", and "somewhat". These operations may have precise definitions, though the definitions can vary considerably between different implementations. "Very", for one example, squares membership functions; since the membership values are always less than 1, this narrows the membership function. "Extremely" cubes the values to give greater narrowing, while "somewhat" broadens the function by taking the square root.

In practice, the fuzzy rule sets usually have several antecedents that are combined using fuzzy operators, such as AND, OR, and NOT, though again the definitions tend to vary: AND, in one popular definition, simply uses the minimum weight of all the antecedents, while OR uses the maximum value. There is also a NOT operator that subtracts a membership function from 1 to give the "complementary" function.

There are several ways to define the result of a rule, but one of the most common and simplest is the "max-min" inference method, in which the output membership function is given the truth value generated by the premise.

Rules can be solved in parallel in hardware, or sequentially in software. The results of all the rules that have fired are "defuzzified" to a crisp value by one of several methods. There are dozens, in theory, each with various advantages or drawbacks.

The "centroid" method is very popular, in which the "center of mass" of the result provides the crisp value. Another approach is the "height" method, which takes the value of the biggest contributor. The centroid method favors the rule with the output of greatest area, while the height method obviously favors the rule with the greatest output value.

The diagram below demonstrates max-min inferencing and centroid defuzzification for a system with input variables "x", "y", and "z" and an output variable "n". Note that "mu" is standard fuzzy-logic nomenclature for "truth value":

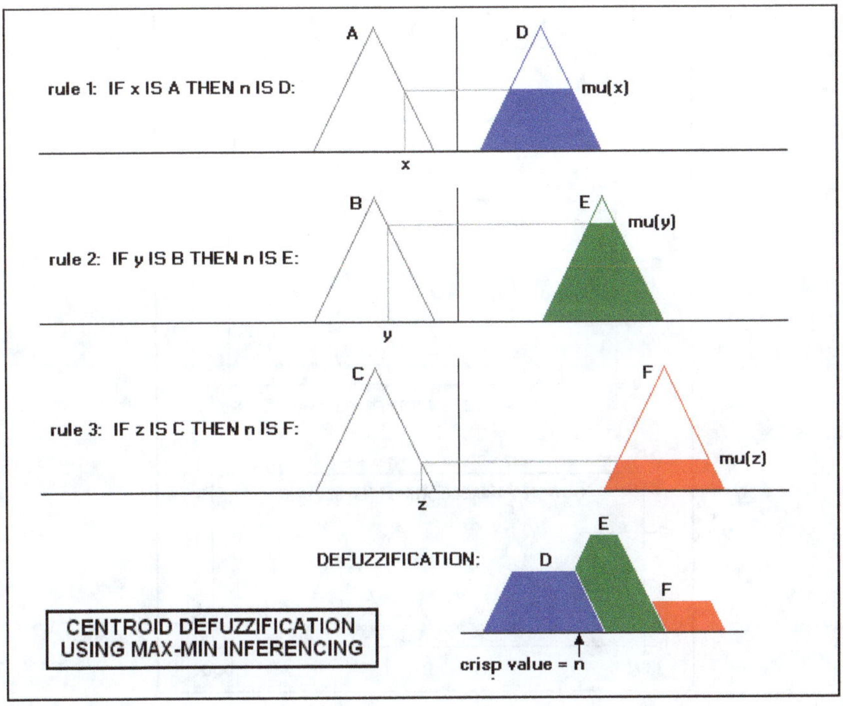

Notice how each rule provides a result as a truth value of a particular membership function for the output variable. In centroid defuzzification the values are OR'd, that is, the maximum value is used and values are not added, and the results are then combined using a centroid calculation.

Fuzzy control system design is based on empirical methods, basically a methodical approach to trial-and-error. The general process is as follows:

- Document the system's operational specifications and inputs and outputs.
- Document the fuzzy sets for the inputs.
- Document the rule set.
- Determine the defuzzification method.
- Run through test suite to validate system, adjust details as required.
- Complete document and release to production.

As a general example, consider the design of a fuzzy controller for a steam turbine. The block diagram of this control system appears as follows.

The input and output variables map into the following fuzzy set:

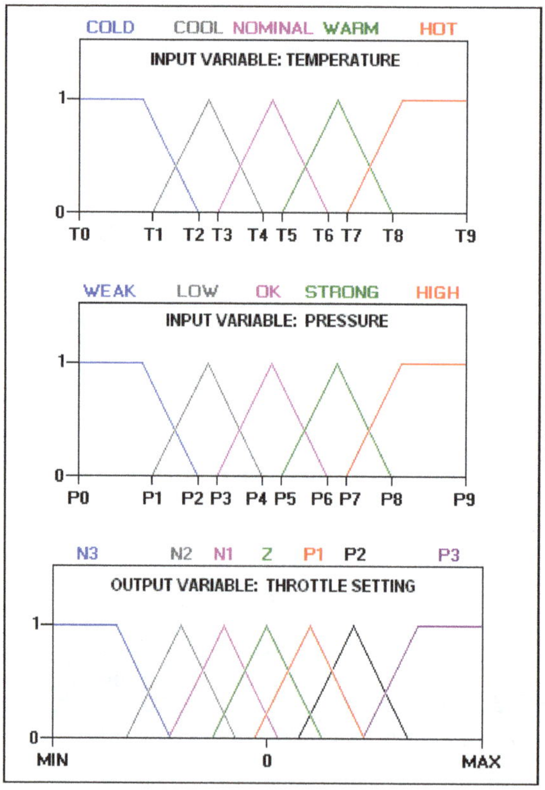

Where:

```
N3: Large negative.
N2: Medium negative.
N1: Small negative.
Z: Zero.
P1: Small positive.
P2: Medium positive.
P3: Large positive.
```

The rule set includes such rules as:

```
rule 1:  IF temperature IS cool AND pressure IS weak,
         THEN throttle is P3.
rule 2:  IF temperature IS cool AND pressure IS low,
         THEN throttle is P2.
rule 3:  IF temperature IS cool AND pressure IS ok,
         THEN throttle is Z.
rule 4:  IF temperature IS cool AND pressure IS strong,
         THEN throttle is N2.
```

In practice, the controller accepts the inputs and maps them into their membership functions and truth values. These mappings are then fed into the rules. If the rule specifies an AND relationship between the mappings of the two input variables, as the examples above do, the minimum of the two is used as the combined truth value; if an OR is specified, the maximum is used. The appropriate output state is selected and assigned a membership value at the truth level of the premise. The truth values are then defuzzified. For an example, assume the temperature is in the "cool" state, and the pressure is in the "low" and "ok" states. The pressure values ensure that only rules 2 and 3 fire:

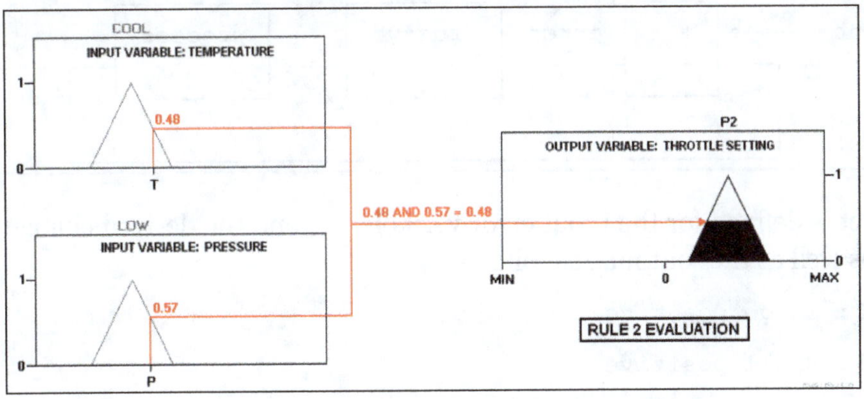

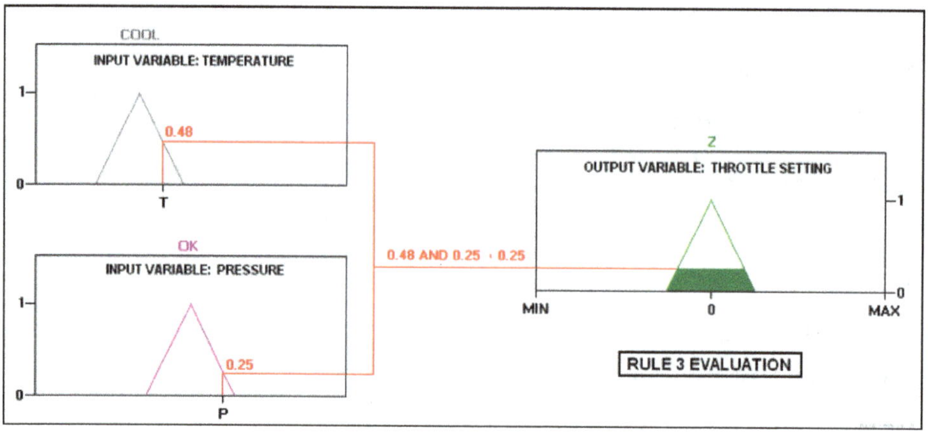

The two outputs are then defuzzified through centroid defuzzification:

```
     |          Z       P2
 1 -+           *        *
     |         * *      * *
     |        *   *    *   *
     |       *     * *      *
     |      *        222222222
     |     *        22222222222
     |    333333332222222222222
     +---333333332222222222222222-->
                     ^
                   +150
```

The output value will adjust the throttle and then the control cycle will begin again to generate the next value.

Building a Fuzzy Controller

Consider implementing with a microcontroller chip a simple feedback controller:

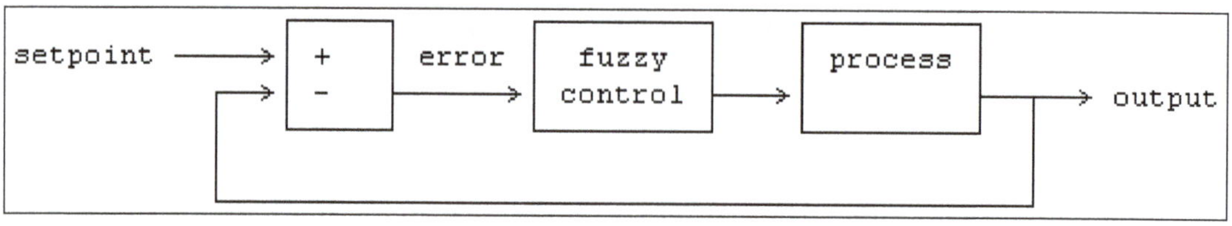

A fuzzy set is defined for the input error variable "e", and the derived change in error, "delta", as well as the "output", as follows:

```
LP: large positive
SP: small positive
```

```
ZE: zero

SN: small negative

LN: large negative
```

If the error ranges from -1 to +1, with the analog-to-digital converter used having a resolution of 0.25, then the input variable's fuzzy set (which, in this case, also applies to the output variable) can be described very simply as a table, with the error / delta / output values in the top row and the truth values for each membership function arranged in rows beneath:

	-1	-0.75	-0.5	-0.25	0	0.25	0.5	0.75	1
mu(LP)	0	0	0	0	0	0	0.3	0.7	1
mu(SP)	0	0	0	0	0.3	0.7	1	0.7	0.3
mu(ZE)	0	0	0.3	0.7	1	0.7	0.3	0	0
mu(SN)	0.3	0.7	1	0.7	0.3	0	0	0	0
mu(LN)	1	0.7	0.3	0	0	0	0	0	0

—or, in graphical form (where each "X" has a value of 0.1):

```
            LN            SN            ZE            SP            LP
        +---------------------------------------------------------------+
        |                                                               |
  -1.0  |  XXXXXXXXXX    XXX            :             :             :   |
  -0.75 |  XXXXXXX       XXXXXXX        :             :             :   |
  -0.5  |  XXX           XXXXXXXXXX     XXX           :             :   |
  -0.25 |  :             XXXXXXX        XXXXXXX       :             :   |
   0.0  |  :             XXX            XXXXXXXXXX    XXX           :   |
   0.25 |  :             :              XXXXXXX       XXXXXXX       :   |
   0.5  |  :             :              XXX           XXXXXXXXXX    XXX |
   0.75 |  :             :              :             XXXXXXX       XXXXXXX |
   1.0  |  :             :              :             XXX           XXXXXXXXXX |
        |                                                               |
        +---------------------------------------------------------------+
```

Suppose this fuzzy system has the following rule base:

```
rule 1: IF e = ZE AND delta = ZE THEN output = ZE

rule 2: IF e = ZE AND delta = SP THEN output = SN

rule 3: IF e = SN AND delta = SN THEN output = LP

rule 4: IF e = LP OR delta = LP THEN output = LN
```

These rules are typical for control applications in that the antecedents consist of the logical combination of the error and error-delta signals, while the consequent is a control command output. The rule outputs can be defuzzified using a discrete centroid computation:

```
SUM( I = 1 TO 4 OF ( mu(I) * output(I) ) ) / SUM( I = 1 TO 4 OF
mu(I) )
```

Now, suppose that at a given time we have:

```
e = 0.25

delta = 0.5
```

Then this gives:

```
         e delta

mu(LP) 0 0.3

mu(SP) 0.7 1

mu(ZE) 0.7 0.3

mu(SN) 0 0

mu(LN) 0 0
```

Plugging this into rule 1 gives:

```
rule 1: IF e = ZE AND delta = ZE THEN output = ZE

      mu(1)  = MIN( 0.7, 0.3 ) = 0.3

output(1) = 0
```

Where:

- mu(1): Truth value of the result membership function for rule 1. In terms of a centroid calculation, this is the "mass" of this result for this discrete case.
- output(1): Value (for rule 1) where the result membership function (ZE) is maximum over the output variable fuzzy set range. That is, in terms of a centroid calculation, the location of the "center of mass" for this individual result. This value is independent of the value of "mu". It simply identifies the location of ZE along the output range.

The other rules give:

```
rule 2: IF e = ZE AND delta = SP THEN output = SN

      mu(2)  = MIN( 0.7, 1 ) = 0.7

output(2) = -0.5


rule 3: IF e = SN AND delta = SN THEN output = LP

      mu(3)  = MIN( 0.0, 0.0 ) = 0

output(3) = 1
```

```
rule 4: IF e = LP OR delta = LP THEN output = LN

        mu(4)   = MAX( 0.0, 0.3 ) = 0.3

output(4) = -1
```

The centroid computation yields:

$$\frac{mu(1)\cdot output(1)+mu(2)\cdot output(2)+mu(3)\cdot output(3)+mu(4)\cdot output(4)}{mu(1)+mu(2)+mu(3)+mu(4)}$$

$$=\frac{(03\cdot 0)+(0.7\cdot -05)+(0\cdot 1)+(0.3\cdot -1)}{03+07+0+0.3}$$

$$=-0.5$$

```
for the final control output. Simple. Of course the hard part is
figuring out what rules actually work correctly in practice.
```

If you have problems figuring out the centroid equation, remember that a centroid is defined by summing all the moments (location times mass) around the center of gravity and equating the sum to zero. So if X_0 is the center of gravity, X_i is the location of each mass, and M_i is each mass, this gives:

$$0=(X_1-X_0)\cdot M_1+(X_2-X_0)\cdot M_2+\ldots+(X_n-X_0)\cdot M_n$$
$$0=(X_1\cdot M_1+X_2\cdot M_2+\ldots+X_n\cdot M_n)-X_0\cdot(M_1+M_2+\ldots+M_n)$$
$$X_0\cdot(M_1+M_2+\ldots+M_n)=X_1\cdot M_1+X_2\cdot M_2+\ldots+X_n\cdot M_n$$
$$X_0=\frac{X_1\cdot M_1+X_2\cdot M_2+\ldots+X_n\cdot M_n}{M_1+M_2+\ldots+M_n}$$

In our example, the values of mu correspond to the masses, and the values of X to location of the masses (mu, however, only 'corresponds to the masses' if the initial 'mass' of the output functions are all the same/equivalent. If they are not the same, i.e. some are narrow triangles, while others maybe wide trapizoids or shouldered triangles, then the mass or area of the output function must be known or calculated. It is this mass that is then scaled by mu and multiplied by its location X_i).

This system can be implemented on a standard microprocessor, but dedicated fuzzy chips are now available. For example, Adaptive Logic INC of San Jose, California, sells a "fuzzy chip", the AL220, that can accept four analog inputs and generate four analog outputs. A block diagram of the chip is shown below:

```
ADC: analog-to-digital converter

DAC: digital-to-analog converter

SH: sample/hold
```

```
                +---------+                                     +-------+
analog --4-->| analog  |                                     | mux / +--4--> analog
   in           |  mux    |                                     |  SH   |         out
                +----+----+                                     +-------+
                     |                                              ^
                     V                                              |
              +-------------+                                   +--+--+
              | ADC / latch |                                   | DAC |
              +------+------+                                   +-----+
                     |                                              ^
                     |                                              |
                     8          +-----------------------------------+
                     |          |                                   |
                     |          V                                   |
                     |   +-----------+       +-------------+        |
                     +-->| fuzzifier |       | defuzzifier +--------+
                         +-----+-----+       +-------------+
                               |                    ^
                               |  +-------------+   |
                               |  |    rule     |   |
                               +->|  processor  +---+
                                  | (50 rules)  |
                                  +------+------+
                                         |
                                  +------+------+
                                  |  parameter  |
                                  |   memory    |
                                  |   256 x 8   |
                                  +-------------+
```

Antilock Brakes

As a first example, consider an anti-lock braking system, directed by a microcontroller chip. The microcontroller has to make decisions based on brake temperature, speed, and other variables in the system.

The variable "temperature" in this system can be subdivided into a range of "states": "cold", "cool", "moderate", "warm", "hot", "very hot". The transition from one state to the next is hard to define.

An arbitrary static threshold might be set to divide "warm" from "hot". For example, at exactly 90 degrees, warm ends and hot begins. But this would result in a discontinuous change when the input value passed over that threshold. The transition wouldn't be smooth, as would be required in braking situations.

The way around this is to make the states *fuzzy*. That is, allow them to change gradually from one state to the next. In order to do this there must be a dynamic relationship established between different factors.

We start by defining the input temperature states using "membership functions":

With this scheme, the input variable's state no longer jumps abruptly from one state to the next. Instead, as the temperature changes, it loses value in one membership function while gaining value in the next. In other words, its ranking in the category of cold decreases as it becomes more highly ranked in the warmer category.

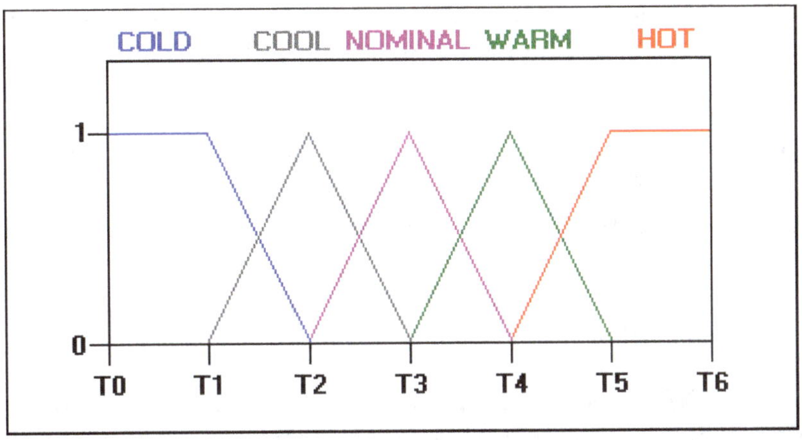

At any sampled timeframe, the "truth value" of the brake temperature will almost always be in some degree part of two membership functions: i.e.: '0.6 nominal and 0.4 warm', or '0.7 nominal and 0.3 cool', and so on.

Adding additional sophistication to this braking system, could be done by additional factors such as traction, speed, inertia, set up in dynamic functions, according to the designed fuzzy system.

Logical Interpretation of Fuzzy Control

In spite of the appearance there are several difficulties to give a rigorous logical interpretation of the *IF-THEN* rules. As an example, interpret a rule as *IF (temperature is "cold") THEN (heater is "high")* by the first order formula *Cold(x)→High(y)* and assume that r is an input such that *Cold(r)* is false. Then the formula *Cold(r)→High(t)* is true for any *t* and therefore any *t* gives a correct control given *r*. A rigorous logical justification of fuzzy control is given in Hájek's book where fuzzy control is represented as a theory of Hájek's basic logic. Also in Gerla 2005 another logical approach to fuzzy control is proposed based on fuzzy logic programming.Indeed, denote by *f* the fuzzy function arising of an IF-THEN systems of rules. Then we can translate this system into a fuzzy program P containing a series of rules whose head is "Good(x,y)". The interpretation of this predicate in the least fuzzy Herbrand model of P coincides with f. This gives further useful tools to fuzzy control.

Fuzzy Qualitative Simulation

Before an Artificial Intelligence system is able to plan the action sequence, some kind of model is needed. For video games, the model is equal to the game rules. From the programming perspective, the game rules are implemented as a Physics engine which accepts an action from a player and calculates, if the action is valid. After the action was executed, the game is in follow up state. If the aim isn't only to play mathematical games but determine the actions for real world applications, the most obvious bottleneck is, that no game rules are available. The first step is to model the domain. System identification can be realized with precise mathematical equations or with Fuzzy rules.

Using Fuzzy logic and ANFIS systems (Adaptive network based fuzzy inference system) for creating the forward model for a domain has many disadvantages. A qualitative simulation isn't able to determine the correct follow up state, but the system will only guess what will happen if the action was taken. The Fuzzy qualitative simulation can't predict the exact numerical values, but it's using imprecise natural language to speculate about the future. It takes the current situation plus the actions from the past and generates the expected follow up state of the game.

The output of the ANFIS system isn't providing correct information, but only a Fuzzy set notation, for example [0,0.2,0.4,0]. After converting the set notation back into numerical values the accuracy get worse. This makes Fuzzy qualitative simulation a bad choice for practical applications.

Fuzzy Markup Language

Fuzzy Markup Language (FML) is a specific purpose markup language based on XML, used for describing the structure and behavior of a fuzzy system independently of the hardware architecture devoted to host and run it.

FML was designed and developed by Giovanni Acampora during his Ph.D. course in Computer Science, at University of Salerno, Italy, in 2004. The original idea inspired Giovanni Acampora to create FML was the necessity of creating a cooperative fuzzy-based framework aimed at automatically controlling a living environment characterized by a plethora of heterogeneous devices whose interactions were devoted to maximize the human comfort under energy saving constraints. This framework represented one of the first concrete examples of Ambient Intelligence. Beyond this pioneering application, the major advantage of using XML to describe a fuzzy system is hardware/software interoperability. Indeed, all that is needed to read an FML file is the appropriate schema for that file, and an FML parser. This markup approach makes it much easier to exchange fuzzy systems between software: for example, a machine learning application could extract fuzzy rules which could then be read directly into a fuzzy inference engine or uploaded into a fuzzy controller. Also, with technologies like XSLT, it is possible to compile the FML into the programming language of your choice, ready for embedding into whatever application you please. As stated by Mike Watts on his popular Computational Intelligence blog:

> "Although Acampora's motivation for developing FML seems to be to develop embedded fuzzy controllers for ambient intelligence applications, FML could be a real boon for developers of fuzzy rule extraction algorithms: from my own experience during my PhD, I know that having to design a file format and implement the appropriate parsers for rule extraction and fuzzy inference engines can be a real pain, taking as much time as implementing the rule extraction

algorithm itself. I would much rather have used something like FML for my work."

A complete overview of FML and related applications can be found in the book titled *On the power of Fuzzy Markup Language* edited by Giovanni Acampora, Chang-Shing Lee, Vincenzo Loia and Mei-Hui Wang, and published by Springer in the series *Studies on Fuzziness and Soft Computing*.

FML at Work: Syntax, Grammar and Hardware Synthesis

FML allows fuzzy systems to be coded through a collection of correlated semantic tags capable of modeling the different components of a classical fuzzy controller such as knowledge base, rule base, fuzzy variables and fuzzy rules. Therefore, the FML tags used to build a fuzzy controller represent the set of lexemes used to create fuzzy expressions. In order to design a well-formed XML-based language, an FML context-free grammar is defined by means of a XML schema which defines name, type and attributes characterized each XML element. However, since an FML program represents only a static view of a fuzzy logic controller, the so-called eXtensible Stylesheet Language Translator (XSLT) is provided to change this static view to a computable version. Indeed, XSLTs modules are able to convert the FML-based fuzzy controller in a general purpose computer language using an XSL file containing the translation description. At this level, the control is executable for the hardware. In short, FML is essentially composed by three layers:

- XML in order to create a new markup language for fuzzy logic control;
- A XML Schema in order to define the legal building blocks;
- A eXtensible Stylesheet Language Transformations (XSLT) in order to convert a fuzzy controller description into a specific programming language.

FML Syntax

FML syntax is composed of XML tags and attributes which describe the different components of a fuzzy logic controller listed below:

- Fuzzy knowledge base;
- Fuzzy rule base;
- Inference engine
- Fuzzification subsystem;
- Defuzzification subsystem.

In detail, the opening tag of each FML program is `<FuzzyController>` which represents the fuzzy controller under modeling. This tag has two attributes: name and ip. The first attribute permits to specify the name of fuzzy controller and ip is used to define the location of controller in a computer network. The fuzzy knowledge base is defined by means of the tag `<KnowledgeBase>` which maintains the set of fuzzy concepts used to model the fuzzy rule base. In order to define the fuzzy concept related controlled system, `<KnowledgeBase>` tag uses a set of nested tags:

- `<FuzzyVariable>` defines the fuzzy concept;
- `<FuzzyTerm>` defines a linguistic term describing the fuzzy concept;
- A set of tags defining a shape of fuzzy sets are related to fuzzy terms.

The attributes of `<FuzzyVariable>` tag are: name, scale, domainLeft, domainRight, type and, for only an output, accumulation, defuzzifier and defaultValue. The name attribute defines the name of fuzzy concept, for instance, temperature; scale is used to define the scale used to measure the fuzzy concept, for instance, Celsius degree; domainLeft and domainRight are used to model the universe of discourse of fuzzy concept, that is, the set of real values related to fuzzy concept, for instance [0°,40°] in the case of Celsius degree; the position of fuzzy concept into rule (consequent part or antecedent part) is defined by type attribute (input/output); accumulation attribute defines the method of accumulation that is a method that permits the combination of results of a variable of each rule in a final result; defuzzifier attribute defines the method used to execute the conversion from a fuzzy set, obtained after aggregation process, into a numerical value to give it in output to system; defaultValue attribute defines a real value used only when no rule has fired for the variable at issue. As for tag `<FuzzyTerm>`, it uses two attributes: name used to identify the linguistic value associate with fuzzy concept and complement, a boolean attribute that defines, if it is true, it is necessary to consider the complement of membership function defined by given parameters. Fuzzy shape tags, used to complete the definition of fuzzy concept, are:

- `<TRIANGULARSHAPE>`
- `<RIGHTLINEARSHAPE>`
- `<LEFTLINEARSHAPE>`
- `<PISHAPE>`
- `<GAUSSIANSHAPE>`
- `<RIGHTGAUSSIANSHAPE>`
- `<LEFTGAUSSIANSHAPE>`
- `<TRAPEZOIDSHAPE>`
- `<SSHAPE>`

- <ZSHAPE>
- <RECTANGULARSHAPE>
- <SINGLETONSHAPE>

Every shaping tag uses a set of attributes which defines the real outline of corresponding fuzzy set. The number of these attributes depends on the chosen fuzzy set shape.

In order to make an example, consider the *Tipper Inference System* described in Mathwork Matlab Fuzzy Logic Toolbox Tutorial. This Mamdani system is used to regulate the tipping in, for example, a restaurant. It has got two variables in input (*food* and *service*) and one in output (*tip*). FML code for modeling part of knowledge base of this fuzzy system containing variables *food* and *tip* is shown below.

```
<?xml version="1.0" encoding="UTF-8"?>
<FuzzyController name="newSystem" ip="127.0.0.1">
    <KnowledgeBase>
        <FuzzyVariable name="food" domainleft="0.0" domainright="10.0"
scale="" type="input">
            <FuzzyTerm name="delicious" complement="false">
                <LeftLinearShape Param1="5.5" Param2="10.0"/>
            </FuzzyTerm>
            <FuzzyTerm name="rancid" complement="false">
             <TriangularShape Param1="0.0" Param2="2.0" Param3="5.5"/>
            </FuzzyTerm>
        </FuzzyVariable>
        ...........
       <FuzzyVariable  name="tip"  domainleft="0.0"  domainright="20.0"
scale="Euro" defaultValue="0.0" defuzzifier="COG"
                      accumulation="MAX"  type="output">
            <FuzzyTerm name="average" complement="false">
           <TriangularShape Param1="5.0" Param2="10.0" Param3="15.0"/>
            </FuzzyTerm>
            <FuzzyTerm name="cheap" complement="false">
            <TriangularShape Param1="0.0" Param2="5.0" Param3="10.0"/>
            </FuzzyTerm>
            <FuzzyTerm name="generous" complement="false">
```

```
          <TriangularShape Param1="10.0" Param2="15.0" Param3="20.0"/>
           </FuzzyTerm>
        </FuzzyVariable>
    <KnowledgeBase>
    ............
</FuzzyController>
```

A special tag that can furthermore be used to define a fuzzy shape is `<UserShape>`. This tag is used to customize fuzzy shape (custom shape). The custom shape modeling is performed via a set of `<Point>` tags that lists the extreme points of geometric area defining the custom fuzzy shape. Obviously, the attributes used in `<Point>` tag are x and y coordinates. As for rule base component, FML allows to define a set of rule bases, each one of them describes a different behavior of system. The root of each rule base is modeled by `<RuleBase>` tag which defines a fuzzy rule set.The `<RuleBase>` tag uses five attributes: *name, type, activationMethod, andMethod* and *orMethod.* Obviously, the *name* attribute uniquely identifies the rule base. The *type* attribute permits to specify the kind of fuzzy controller (Mamdani or TSK) respect to the rule base at issue. The *activationMethod* attribute defines the method used to implication process; the *andMethod* and *orMethod* attribute define, respectively, the *and* and *or* algorithm to use by default. In order to define the single rule the `<Rule>` tag is used. The attributes used by the `<Rule>` tag are: *name, connector, operator* and *weight.* The *name* attribute permits to identify the rule; *connector* is used to define the logical operator used to connect the different clauses in antecedent part (and/or); *operator* defines the algorithm to use for chosen connector; *weight* defines the importance of rule during inference engine step. The definition of antecedent and consequent rule part is obtained by using `<Antecedent> and <Consequent> tags.` `<Clause>` tag is used to model the fuzzy clauses in antecedent and consequent part. This tag use the attribute *modifier* to describe a modification to term used in the clause. The possible values for this attribute are: *above, below, extremely, intensify, more or less, norm, not, plus, slightly, somewhat, very, none.* To complete the definition of fuzzy clause the nested `<Variable>` and `<Term>` tags have to be used. A sequence of `<Rule>` tags realizes a fuzzy rule base.

As example, consider a Mamdani rule composed by *(food is rancid) OR (service is very poor)* as antecedent and *tip is cheap* as consequent. The antecedent part is formed by two clauses: *(food is rancid)* and *(service is poor).* The first antecedent clause uses *food* as variable and *rancid* as fuzzy term, whereas, the second antecedent clause uses *service* as a variable, *poor* as fuzzy term and *very* as modifier; the consequent clause uses *tip* as a fuzzy variable and *cheap* as a fuzzy term. The complete rule is:

IF (food is rancid) OR (service is very poor) THEN (tip is cheap).

Let us see how FML defines a rule base with this rule.

```
<RuleBase name="Rulebase1" activationMethod="MIN" andMethod="MIN" or-
Method="MAX" type="mamdani">
      <Rule name="reg1" connector="or" operator="MAX" weight="1.0">
           <Antecedent>
                <Clause>
                     <Variable>food</Variable>
                     <Term>rancid</Term>
                </Clause>
                <Clause modifier="very">
                     <Variable>service</Variable>
                     <Term>poor</Term>
                </Clause>
           </Antecedent>
           <Consequent>
                <Clause>
                     <Variable>tip</Variable>
                     <Term>cheap</Term>
                </Clause>
           </Consequent>
      </Rule>
      ............
</RuleBase>
```

Now, let us see a Takagi-Sugeno-Kang system that regulates the same issue. The most important difference with Mamdani system is the definition of a different output variable *tip*. The `<TSKVariable>` tag is used to define an output variable that can be used in a rule of a Tsk system. This tag has the same attributes of a Mamdani output variable except for the *domainleft* and *domainright* attribute because a variable of this kind (called tsk-variable) hasn't a universe of discourse. The nested `<TSKTerm>` tag represents a linear function and so it is completely different from `<FuzzyTerm>`. The `<TSKValue>` tag is used to define the coefficients of linear function. The following crunch of FML code shows the definition of output variable *tip* in a Tsk system.

```
<?xml version="1.0" encoding="UTF-8"?>
<FuzzyController name="newSystem" ip="127.0.0.1">
    <KnowledgeBase>
```

```
      .......
          <TSKVariable name="tip" scale="null" accumulation="MAX" de-
fuzzifier="WA" type="output">
            <TSKTerm name="average" order="0">
                <TSKValue>1.6</TSKValue>
            </TSKTerm>
            <TSKTerm name="cheap" order="1">
                <TSKValue>1.9</TSKValue>
                <TSKValue>5.6</TSKValue>
                <TSKValue>6.0</TSKValue>
            </TSKTerm>
            <TSKTerm name="generous" order="1">
                <TSKValue>0.6</TSKValue>
                <TSKValue>1.3</TSKValue>
                <TSKValue>1.0</TSKValue>
            </TSKTerm>
        </TSKVariable>
    <KnowledgeBase>
    ..........
</FuzzyController>
```

The FML definition of rule base component in a Tsk system doesn't change a lot. The only different thing is that the `<Clause>` tag doesn't have the modifier attribute.

As example, consider a tsk rule composed by *(food is rancid) OR (service is very poor)* as antecedent and, as consequent, *tip=1.9+5.6*food+6.0*service* that can be written as *tip is cheap* in an implicitly way. So the rule can be written in this way:

IF (food is rancid) OR (service is very poor) THEN (tip is cheap).

Let us see how FML defines a rule base with this rule.

```
<RuleBase name="Rulebase1" activationMethod="MIN" andMethod="MIN" or-
Method="MAX" type="tsk">
<Rule name="reg1" connector="or" operator="MAX" weight="1.0">
            <Antecedent>
                <Clause>
                    <Variable>food</Variable>
```

```
                    <Term>rancid</Term>
                </Clause>
                <Clause>
                    <Variable>service</Variable>
                    <Term>poor</Term>
                </Clause>
            </Antecedent>
            <Consequent>
                <Clause>
                    <Variable>tip</Variable>
                    <Term>cheap</Term>
                </Clause>
            </Consequent>
        </Rule>
      ............
</RuleBase>
```

FML Grammar

The FML tags used to build a fuzzy controller represent the set of lexemes used to create fuzzy expressions. However, in order to realize a well-formed XML-based language, an FML context-free grammar is necessary and described in the following. The FML context-free grammar is modeled by XML file in the form of a XML Schema Document (XSD) which expresses the set of rules to which a document must conform in order to be considered a *valid* FML document. Based on the previous definition, a portion of the FML XSD regarding the knowledge base definition is given below.

```
<?xml version="1.0" encoding="UTF-8"?>
<xs:schema xmlns:xs="http://www.w3.org/2001/XMLSchema">
  ........
      <xs:complexType name="KnowledgeBaseType">
            <xs:sequence>
               <xs:choice minOccurs="0" maxOccurs="unbounded">
                    <xs:element name="FuzzyVariable" type="FuzzyVari-
ableType"/>
                    <xs:element name="TSKVariable"        type="TSK-
```

```
VariableType"/>
                </xs:choice>
            </xs:sequence>
        </xs:complexType>
        <xs:complexType name="FuzzyVariableType">
            <xs:sequence>
                <xs:element name="FuzzyTerm" type="FuzzyTermType"
maxOccurs="unbounded"/>
            </xs:sequence>
            <xs:attribute name="name" type="xs:string" use="re-
quired"/>
            <xs:attribute name="defuzzifier" default="COG">
                <xs:simpleType>
                    <xs:restriction base="xs:string">
                        <xs:pattern value="MM|COG|COA|WA|Cus-
tom"/>
                    </xs:restriction>
                </xs:simpleType>
            </xs:attribute>
            <xs:attribute name="accumulation" default="MAX">
                <xs:simpleType>
                    <xs:restriction base="xs:string">
                        <xs:pattern value="MAX|SUM"/>
                    </xs:restriction>
                </xs:simpleType>
            </xs:attribute>
            <xs:attribute name="scale" type="xs:string" />
            <xs:attribute name="domainleft" type="xs:float" use="re-
quired"/>
            <xs:attribute name="domainright" type="xs:float" use="re-
quired"/>
            <xs:attribute name="defaultValue" type="xs:float" de-
fault="0"/>
            <xs:attribute name="type" default="input">
```

```
                <xs:simpleType>
                      <xs:restriction base="xs:string">
                            <xs:pattern value="input|output"/>
                      </xs:restriction>
                </xs:simpleType>
          </xs:attribute>
    </xs:complexType>
    <xs:complexType name="FuzzyTermType">
          <xs:choice>
                <xs:element name="RightLinearShape" type="TwoParam-
Type"/>
                <xs:element name="LeftLinearShape" type="TwoParam-
Type"/>
                <xs:element name="PIShape" type="TwoParamType"/>
                <xs:element name="TriangularShape" type="ThreePa-
ramType"/>
                <xs:element name="GaussianShape" type="TwoParam-
Type"/>
                <xs:element name="RightGaussianShape" type="Two-
ParamType"/>
                <xs:element name="LeftGaussianShape" type="TwoParam-
Type"/>
                <xs:element name="TrapezoidShape" type="FourParam-
Type"/>
                <xs:element name="SingletonShape" type="OneParam-
Type"/>
                <xs:element name="RectangularShape" type="TwoParam-
Type"/>
                <xs:element name="ZShape" type="TwoParamType"/>
                <xs:element name="SShape" type="TwoParamType"/>
                <xs:element name="UserShape" type="UserShapeType"/>
          </xs:choice>
            <xs:complexType name="TwoParamType">
          <xs:attribute name="Param1" type="xs:float" use="re-
quired"/>
```

```
            <xs:attribute   name="Param2"   type="xs:float"   use="re-
quired"/>
      </xs:complexType>
      <xs:complexType name="ThreeParamType">
            <xs:attribute   name="Param1"   type="xs:float"   use="re-
quired"/>
            <xs:attribute   name="Param2"   type="xs:float"   use="re-
quired"/>
            <xs:attribute   name="Param3"   type="xs:float"   use="re-
quired"/>
      </xs:complexType>
      <xs:complexType name="FourParamType">
            <xs:attribute   name="Param1"   type="xs:float"   use="re-
quired"/>
            <xs:attribute   name="Param2"   type="xs:float"   use="re-
quired"/>
            <xs:attribute   name="Param3"   type="xs:float"   use="re-
quired"/>
            <xs:attribute   name="Param4"   type="xs:float"   use="re-
quired"/>
      </xs:complexType>
      <xs:complexType name="UserShapeType">
            <xs:sequence>
                  <xs:element  name="Point"  type="PointType"  minOc-
curs="2" maxOccurs="unbounded"/>
            </xs:sequence>
      </xs:complexType>
      <xs:complexType name="PointType">
            <xs:attribute name="x" type="xs:float" use="required"/>
            <xs:attribute name="y" type="xs:float" use="required"/>
      </xs:complexType>
      <xs:complexType name="RuleBaseType">
            <xs:attribute   name="name"   type="xs:string"   use="re-
quired"/>
            <xs:attribute name="activationMethod" default="MIN">
```

```
            <xs:simpleType>
                  <xs:restriction base="xs:string">
                        <xs:pattern value="PROD|MIN"/>
                  </xs:restriction>
            </xs:simpleType>
      </xs:attribute>
      <xs:attribute name="andMethod" default="MIN">
            <xs:simpleType>
                  <xs:restriction base="xs:string">
                        <xs:pattern value="PROD|MIN"/>
                  </xs:restriction>
            </xs:simpleType>
      </xs:attribute>
      <xs:attribute name="orMethod" default="MAX">
            <xs:simpleType>
                  <xs:restriction base="xs:string">
                        <xs:pattern value="PROBOR|MAX"/>
                  </xs:restriction>
            </xs:simpleType>
      </xs:attribute>
      <xs:attribute name="type" use="required">
            <xs:simpleType>
                        <xs:restriction base="xs:string">
                              <xs:pattern                         val-
ue="TSK|Tsk|tsk|Mamdani|mamdani"/>
                        </xs:restriction>
            </xs:simpleType>
      </xs:attribute>
</xs:complexType>
<xs:complexType name="MamdaniRuleBaseType">
      <xs:complexContent>
      <xs:extension base="RuleBaseType">
```

```
                    <xs:sequence>
                          <xs:element name="Rule" type="Mamdani-
FuzzyRuleType" minOccurs="0" maxOccurs="unbounded"/>
                    </xs:sequence>
              </xs:extension>
        </xs:complexContent>
  </xs:complexType>
  <xs:complexType name="AntecedentType">
        <xs:sequence>
              <xs:element name="Clause" type="ClauseType" maxOc-
curs="unbounded"/>
        </xs:sequence>
  </xs:complexType>
  <xs:complexType name="MamdaniConsequentType">
        <xs:sequence>
              <xs:element name="Clause" type="ClauseType" maxOc-
curs="unbounded"/>
        </xs:sequence>
  </xs:complexType>
  <xs:complexType name="ClauseType">
        <xs:sequence>
              <xs:element name="Variable">
                    <xs:simpleType>
                          <xs:restriction base="xs:string">
                                <xs:whiteSpace value="collapse"/>
                                <xs:pattern value="(([A-Z])|([a-
z]))+([A-Z]|[a-z]|[0-9])*"/>
                          </xs:restriction>
                    </xs:simpleType>
              </xs:element>
              <xs:element name="Term" type="xs:string">
              </xs:element>
```

```
        </xs:sequence>
        <xs:attribute name="modifier" use="optional">
            <xs:simpleType>
                <xs:restriction base="xs:string">
                    <xs:pattern
                                        value="above|below|extreme-
ly|intensify|more_or_less|norm|not|plus|slightly|somewhat|very"/>
                </xs:restriction>
            </xs:simpleType>
        </xs:attribute>
    </xs:complexType>
    ..........
</xs:schema>
```

FML Synthesis

Since an FML program realizes only a static view of a fuzzy system, the so-called eXtensible Stylesheet Language Translator (XSLT) is provided to change this static view to a computable version. In particular, the XSLT technology is used convert a fuzzy controller description into a general-purpose computer language to be computed on several hardware platforms. Currently, a XSLT converting FML program in runnable Java code has been implemented. In this way, thanks to the transparency capabilities provided by Java virtual machines, it is possible to obtain a fuzzy controller modeled in high level way by means of FML and runnable on a plethora of hardware architectures through Java technologies. However, XSLT can be also used for converting FML programs in legacy languages related to a particular hardware or in other general purpose languages.

EXPERT SYSTEM

A fuzzy expert system is an expert system that uses a collection of fuzzy membership functions and rules, instead of Boolean logic, to reason about data. The rules in a fuzzy expert system are usually of a form similar to the following:

if x is low and y is high then z = medium

where, x and y are input variables (names for know data values), z is an output variable (a name for a data value to be computed), low is a membership function (fuzzy subset) defined on x, high is a membership function defined on y, and medium is a membership function defined on z. The antecedent (the rule's premise) describes to what degree the rule applies, while the conclusion (the rule's consequent) assigns a membership function to each of one or more output variables. Most tools for working with fuzzy expert systems allow more than one conclusion per rule. The set of rules in a fuzzy expert system is known as the rulebase or knowledge base.

The general inference process proceeds in three (or four) steps:

1. Under FUZZIFICATION, the membership functions defined on the input variables are applied to their actual values, to determine the degree of truth for each rule premise.

2. Under INFERENCE, the truth value for the premise of each rule is computed, and applied to the conclusion part of each rule. This results in one fuzzy subset to be assigned to each output variable for each rule. Usually only MIN or PRODUCT are used as inference rules. In MIN inferencing, the output membership function is clipped off at a height corresponding to the rule premise's computed degree of truth (fuzzy logic AND). In PRODUCT inferencing, the output membership function is scaled by the rule premise's computed degree of truth.

3. Under COMPOSITION, all of the fuzzy subsets assigned to each output variable are combined together to form a single fuzzy subset for each output variable. Again, usually MAX or SUM are used. In MAX composition, the combined output fuzzy subset is constructed by taking the pointwise maximum over all of the fuzzy subsets assigned tovariable by the inference rule (fuzzy logic OR). In SUM composition, the combined output fuzzy subset is constructed by taking the pointwise sum over all of the fuzzy subsets assigned to the output variable by the inference rule.

4. Finally is the (optional) DEFUZZIFICATION, which is used when it is useful to convert the fuzzy output set to a crisp number. There are more defuzzification methods than you can shake a stick at (at least 30). Two of the more common techniques are the CENTROID and MAXIMUM methods. In the CENTROID method, the crisp value of the output variable is computed by finding the variable value of the center of gravity of the membership function for the fuzzy value. In the MAXIMUM method, one of the variable values at which the fuzzy subset has its maximum truth value is chosen as the crisp value for the output variable.

Extended Example:

Assume that the variables x, y, and z all take on values in the interval [0,10], and that the following membership functions and rules are defined:

$$low(t) = 1 - (t / 10)$$

$$high(t) = t / 10$$

- Rule 1: If x is low and y is low then z is high.
- Rule 2: If x is low and y is high then z is low.
- Rule 3: If x is high and y is low then z is low.
- Rule 4: If x is high and y is high then z is high.

Notice that instead of assigning a single value to the output variable z, each rule assigns an entire fuzzy subset (low or high).

In the fuzzification subprocess, the membership functions defined on the input variables are applied to their actual values, to determine the degree of truth for each rule premise. The degree of truth for a rule's premise is sometimes referred to as its ALPHA. If a rule's premise has a nonzero degree of truth (if the rule applies at all) then the rule is said to FIRE. For example:

x	y	low(x)	high(x)	low(y)	high(y)	alpha1	alpha2	alpha3	alpha4
0.0	0.0	1.0	0.0	1.0	0.0	1.0	0.0	0.0	0.0
0.0	3.2	1.0	0.0	0.68	0.32	0.68	0.32	0.0	0.0
0.0	6.1	1.0	0.0	0.39	0.61	0.39	0.61	0.0	0.0
0.0	10.0	1.0	0.0	0.0	1.0	0.0	1.0	0.0	0.0
3.2	0.0	0.68	0.32	1.0	0.0	0.68	0.0	0.32	0.0
6.1	0.0	0.39	0.61	1.0	0.0	0.39	0.0	0.61	0.0
10.0	0.0	0.0	1.0	1.0	0.0	0.0	0.0	1.0	0.0
3.2	3.1	0.68	0.32	0.69	0.31	0.68	0.31	0.32	0.31
3.2	3.3	0.68	0.32	0.67	0.33	0.67	0.33	0.32	0.32
10.0	10.0	0.0	1.0	0.0	1.0	0.0	0	0.0	1.0

In the inference subprocess, the truth value for the premise of each rule is computed, and applied to the conclusion part of each rule. This results in one fuzzy subset to be assigned to each output variable for each rule.

MIN and PRODUCT are two INFERENCE METHODS or INFERENCE RULES. In MIN inferencing, the output membership function is clipped off at a height corresponding to the rule premise's computed degree of truth. This corresponds to the traditional interpretation of the fuzzy logic AND operation. In PRODUCT inferencing, the output membership function is scaled by the rule premise's computed degree of truth.

For example, let's look at rule 1 for x = 0.0 and y = 3.2. As shown in the table, the premise degree of truth works out to 0.68. For this rule, MIN inferencing will assign z the fuzzy subset defined by the membership function:

rule1(z) = { z / 10, if z <= 6.8

0.68, if z >= 6.8 }

For the same conditions, PRODUCT inferencing will assign z the fuzzy subset defined by the membership function:

rule1(z) = 0.68 * high(z)

= 0.068 * z

In the composition subprocess, all of the fuzzy subsets assigned to each output variable are combined together to form a single fuzzy subset for each output variable.

MAX composition and SUM composition are two COMPOSITION RULES. In MAX composition, the combined output fuzzy subset is constructed by taking the pointwise maximum over all of the fuzzy subsets assigned to the output variable by the inference rule. In SUM composition, the combined output fuzzy subset is constructed by taking the pointwise sum over all of the fuzzy subsets assigned to the output variable by the inference rule. Note that this can result in truth values greater than one! For this reason, SUM composition is only used when it will be followed by a defuzzification method, such as the CENTROID method, that doesn't have a problem with this odd case. Otherwise SUM composition can be combined with normalization and is therefore a general purpose method again. For example, assume x = 0.0 and y = 3.2. MIN inferencing would assign the following four fuzzy subsets to z:

rule1(z) = { z / 10, if z <= 6.8
0.68, if z >= 6.8 }

rule2(z) = { 0.32, if z <= 6.8
1 - z / 10, if z >= 6.8 }

rule3(z) = 0.0

rule4(z) = 0.0

MAX composition would result in the fuzzy subset:

fuzzy(z) = { 0.32, if z <= 3.2
z / 10, if 3.2 <= z <= 6.8
0.68, if z >= 6.8 }

PRODUCT inferencing would assign the following four fuzzy subsets to z:

rule1(z) = 0.068 * z

rule2(z) = 0.32 - 0.032 * z

rule3(z) = 0.0

rule4(z) = 0.0

SUM composition would result in the fuzzy subset:

fuzzy(z) = 0.32 + 0.036 * z

Sometimes it is useful to just examine the fuzzy subsets that are the result of the composition process, but more often, this FUZZY VALUE needs to be converted to a single number - a CRISP VALUE. This is what the defuzzification subprocess does.

There are more defuzzification methods than you can shake a stick at. A couple of years ago, Mizumoto did a short paper that compared about ten defuzzification methods. Two of the more common techniques are the CENTROID and MAXIMUM methods. In the CENTROID method, the crisp value of the output variable is computed by finding the variable value of the center of gravity of the membership function for the fuzzy value. In the MAXIMUM method, one of the variable values at which the fuzzy subset has its maximum truth value is chosen as the crisp value for the output variable. There are several variations of the MAXIMUM method that differ only in what they do when there is more than one variable value at which this maximum truth value occurs. One of these, the AVERAGE-OF-MAXIMA method, returns the average of the variable values at which the maximum truth value occurs.

For example, go back to our previous examples. Using MAX-MIN inferencing and AVERAGE-OF-MAXIMA defuzzification results in a crisp value of 8.4 for z. Using PRODUCT-SUM inferencing and CENTROID defuzzification results in a crisp value of 5.6 for z, as follows.

Earlier on in the FAQ, we state that all variables (including z) take on values in the range [0, 10]. To compute the centroid of the function f(x), you divide the moment of the function by the area of the function. To compute the moment of f(x), you compute the integral of x*f(x) dx, and to compute the area of f(x), you compute the integral of f(x) dx. In this case, we would compute the area as integral from 0 to 10 of (0.32+0.036*z) dz, which is:

(0.32 * 10 + 0.018*100)

= (3.2 + 1.8)

= 5.0

and the moment as the integral from 0 to 10 of (0.32*z+0.036*z*z) dz, which is:

(0.16 * 10 * 10 + 0.012 * 10 * 10 * 10)

= (16 + 12)

= 28

Finally, the centroid is 28/5 or 5.6.

FUZZY LOGIC IN MOBILE ROBOT NAVIGATION

An autonomous robot is a programmable and multi-functional machine, able to extract information from its surrounding using different kinds of sensors to plan and execute collision free motions within its environment without human intervention. Navigation is a crucial issue for robots that claim to be mobile. A navigation system can be divided into two layers: High level global planning and Low-level reactive control. In high-level planning, a prior knowledge of environment is available and the robot workspace is completely or partially known. Using the world model, the global planner can determine the robot motion direction and generates minimum-cost paths towards the target in the presence of complex obstacles. However, since it is not capable of changing the motion direction in presence of unforeseen or moving obstacles, it fails to reach target. In contrast, in low-level reactive control, the robot work space is unknown and dynamic. It generates control commands based on perception-action configuration, which the robot uses current sensory information to take appropriate actions without planning process. Thus, it has a quick response in reacting to unforeseen obstacles and uncertainties with changing the motion direction.

Several Artificial intelligence techniques such as reinforcement learning, neural networks, fuzzy logic and genetic algorithms, can be applied for the reactive navigation of mobile robots to improve their performance. Amongst the techniques ability of fuzzy logic to represent linguistic terms and reliable decision making in spite of uncertainty and imprecise information makes it a useful tool in control systems.

Fuzzy control systems are rule-based or knowledge-based systems containing a collection of fuzzy IF-THEN rules based on the domain knowledge or human experts. The simplicity of fuzzy rule-based systems, capability to perform a wide variety tasks without explicit computations and measurements make it extensively popular among the scientists and researcher.

Robust and reliable navigation in dynamic or unknown environment relies on ability of the robots in moving among unknown obstacles without collision and fast reaction to uncertainties. It is highly desirable to develop these tasks using a technique which utilize human reasoning and decision making. Fuzzy logic provides a means to capture the human mind's expertise. It utilizes this heuristic knowledge for representing and accomplishment of a methodology to develop perception-action based strategies for mobile robots navigation. Furthermore, the methodology of the FLC is very helpful dealing with uncertainties in real world and accurate model of the environment is not absolutely required for navigation. Therefore, based on a simple design, easy implementation and robustness properties of FLC, many approaches were developed to solve mobile robot navigation problem in target tracking, path tracking, obstacle avoidance, behavior coordination, environment modelling, and layer integration.

Fuzzy Logic for Path Tracking

Path tracking is a crucial function for autonomous mobile robots to navigate along a desired path. This task includes tracking of previously computed paths using a path planner, a defined path by human operator, tracking of walls, road edges, and other natural features in the robot workspace. It involves real-time perception of the environment to determine the position and orientation of the robot with respect to the desired path. For example in figure, if the robot is misplaced, the controller task is to steer it back on course and minimize the orientation error ($\Delta\varphi$) and the position error (Δx). Path tracking difficulties in dealing with imprecise or incomplete perception of environment, representation of inaccuracy in measurements, sensor fusion and compliance with the kinematic limits of the vehicle motivated many researchers to use fuzzy control techniques for path tracking.

Ollero et al. developed a new fuzzy path-tracking method by combining fuzzy logic with the geometric pure-pursuit and the generalized predictive control techniques. Fuzzy logic is applied to supervise path trackers. Input of the fuzzy is the current state of the robot to the path to generate the appropriate steering angle. A new approach proposed by Braunstingl et al. to solve the wall following of mobile robots based on the concept of general perception. To construct a general perception of the surroundings from the measuring data provided by all the sensors and representing, a perception vector is assigned to each ultrasonic sensor. All these vectors adding together then combine into a single vector of general perception. A fuzzy controller then uses the perception information to guide the robot along arbitrary walls and obstacles. Sanchez et al. proposed a fuzzy control system for path tracking of an autonomous vehicle in outdoor environment. The fuzzy controller is used to generate steering and velocity required to track the path using the data collected from experiments of driving the vehicle by a human. Bento et al. implemented a path-tracking method by means of fuzzy logic for a Wheeled Mobile Robot. Input variables of the fuzzy controller are position and orientation of the robot with respect to the path. Output variables are linear velocity and angular velocity. Hajjaji and Bentalba have designed a fuzzy controller for path tracking control of vehicles using its nonlinear dynamics model. A Takagi–Sugeno (T–S) fuzzy model presents the nonlinear model of the vehicle. Then a model-based fuzzy controller is developed based on the T–S fuzzy model. A wall-following robot presented by Peri & Simon which the robot's motion is controlled by a fuzzy controller to drive it along a predefined path. Antonelli et al. address a path tracking approach based on a fuzzy-logic set of rules which emulates the human driving behavior. The fuzzy system input is represented by approximate information concerning the knowledge of the curvature of the desired path ahead the vehicle and the distance between the next bend and the vehicle. The output is the maximum value of the linear velocity needed to attain by the vehicle in order to safely drive on the path. Yu et al. used Taguchi method to design an optimal fuzzy logic controller for trajectory tracking of a wheeled mobile robot. Recently, Xiong and Qu developed a method for intelligent vehicles' path tracking with two fuzzy controller combinations which controls vehicle direction and a preview fuzzy control method presented by Liao et al. for path tracking of intelligent vehicle. The vehicle speed and direction are adjusted by fuzzy control according to future path information and present path information respectively.

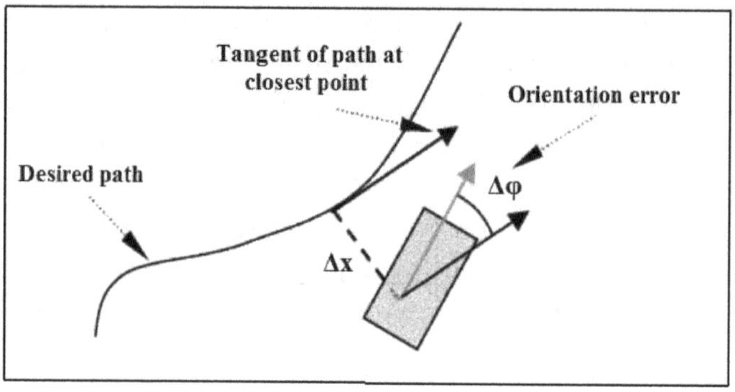

Typical control input variables for path tracking.

Obstacle Avoidance using Fuzzy Logic

Ability of a robot to avoid collision with unforeseen or dynamic obstacles while it is moving towards a target or tracking a path is a vital task in autonomous navigation. Navigation strategies can be classified to global path planning and local path planning. In global path planning, information about the obstacles and a global model of environment is available which mostly Configuration space, Road map, Voronoi diagram and Potential field techniques are used to plan obstacle-free path towards a target. However, in real world a reliable map of obstacle, accurate model of environment and precise sensory data is unavailable due to uncertainties of the environment. While the computed path may remain valid but to response the unforeseen or dynamic obstacles, it is necessary for the robot to alter its path online. In such situations, Fuzzy logic can provide robust and reliable methodologies dealing with the imprecise input with low computational complexity. Different obstacle avoidance approaches were developed during past decades which proposed effective solution to the navigation problems in unknown and dynamic environments.

Chee et al. presented a two-layer fuzzy inference system in which the first layer fuses the sensor readings. The left and right clearances of the robot were found as outputs of the first-layered fuzzy system. The outputs of the first layer together with the goal direction are used as the inputs of the second-layer. Eventually, the final outputs of the controller are the linear velocity and the turning rate of the robot. The second-stage fuzzy inference system employs the collision avoiding, obstacle following and goal tracking behaviors to achieve robust navigation in unknown environments. Dadios and Maravillas proposed and implemented a fuzzy control approach for cooperative soccer micro robots. A planner generates a path to the destination and fuzzy logic control the robot's heading direction to avoid obstacles and other robots while the dynamic position of obstacles, ball and robots are considered. Zavlangas et al. developed a reactive navigation method for omnidirectional mobile robots using fuzzy logic. The fuzzy rule-base generates actuating command to get collision free motions in dynamic environment. The fuzzy logic also provides an adjustable transparent system by a set of learning rules or manually. Seraji and Howard developed a behavior-based navigation method on challenging terrain using fuzzy logic. The navigation strategy is comprised of three behaviors. Local obstacle avoidance behavior is consists of a set of fuzzy logic rule statements which generates the robot's speed based on obstacle distance. Parhi described a control system comprises a fuzzy logic controller and a Petri Net for multi robot navigation. The Fuzzy rules steer the robot according to obstacles distribution or targets position. Since the obstacle's position is not known precisely, to avoid obstacles in a cluttered environment fuzzy logic is a proper technique for this task. Combination of the fuzzy logic controller and a set of collision prevention rules implemented as a Petri Net model embedded in the controller of a mobile robot enable it to avoid obstacles that include other mobile robots. A fuzzy controller designed by Lilly for obstacle avoidance of an autonomous vehicle using negative fuzzy rules. The negative fuzzy rules define a set of actions to be avoided to direct the vehicle to a target in presence of obstacles. Chao et al. eveloped a fuzzy control system for target tracking and obstacle avoidance of a mobile robot. Decision making is handled by the fuzzy control strategy based on the sensed environment using a stereo vision information. A vision- based fuzzy obstacle avoidance proposed for a humanoid robot in. The nearest obstacle to the robot captured by vision system and the difference angle between goal direction and the robot's heading measured by electronic compass are inputs of the fuzzy system to make a decision for appropriate motion of the robot in unknown environment.

Fuzzy Logic for Behavior Coordination

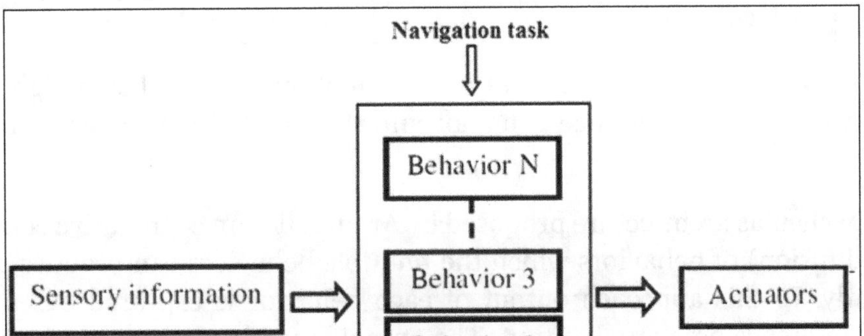

Behavior-based navigation systems overall architecture.

To improve the total performance of a navigation system, complex navigation tasks are broken down into a number of simpler and smaller subsystems (behaviors) which is called behaviorbased system. In a behavior-based system, each behavior receives particular sensory information and transforms them into the predefined response. The behaviors include path tracking, obstacle avoidance, target tracking, goal reaching and etc. Finally, based on command output(s) of an active behavior(s) the robot executes an action.

The problems associated with the behavior-based navigation systems is the behavior coordination or action selection. The multiple behaviors may produce several command outputs simultaneously which may cause the robot move in unintended directions or system fail entirely. Reliable and robust operation of the system relies on the decision about how to integrate high level planning and low level execution behaviors, which behavior should be activated (arbitration) and how output commands should be combined into one command to drive the robot (command fusion). Early solutions were developed based on subsumption architecture and motor schemas.

The subsumption architecture is composed of several layers of task-achieving behaviors. Coordination of behaviors is based on Priority arbitration (Competitive architecture). In Priority-based arbitration only a behavior with the highest priority is selected to be active when multiple conflicting behaviors are trigged and the other are ignored. The subsumption approach is based on a static arbitration policy which means that the robot actions are predefined and fixed in dealing with certain situations. Since the behavior coordination is competitive and based on a fixed arbitration, it may leads to erratic operation under certain situations. For example in coordination of goal reaching and obstacle avoidance behaviors with rules like:

Obstacle avoidance rules	Goal reaching rules
IF Obstacle is left THEN turn right.	IF goal is right THEN turn right.
IF Obstacle is front THEN turn left.	IF goal is left THEN turn left.

When an obstacle is detected in front of the robot and the goal is at right, the priority is with Obstacle avoidance behavior and the robot turns left while the goal is at right.

The motor schemas architecture proposed by Arkin relies on cooperative coordination (command fusion) of behaviors which the multiple behaviors can produce an output concurrently. In this approach output of each behavior is captured based on their particular influence on overall output. The outputs are blended to vote for or against an action. For example in potential fields the outputs are in the vector form. These outputs are combined and the overall response of the system is achieved by the vector summation. This approach also may lead to conflicting actions or poor performance in certain circumstances. However, fuzzy logic provides a useful mechanism for command fusion coordination and also arbitration fusion coordination. The main fuzzy logic advantages are: i) it can be used for dynamic arbitration which behavior selection is according to the robot's current perceptual state, ii) it allows for easy combination and concurrent execution of various behaviors. A variety of approaches have been developed inspired by the success of fuzzy logic to deal with the behavior coordination limitations.

Leyden designed a fuzzy logic based navigation system to overcome the subsumption control problem. The proposed system is consists of two behaviors. Output of each behavior is a fuzzy set which are combined using a command fusion process to produce a single fuzzy set. Then, the fuzzy set is defuzzified to make a crisp output. Fatmi et al. proposed a two layered behavior coordination approach for behavior design and action coordination using fuzzy logic. The first layer is consists of primitive basic behaviors and the second layer is responsible for decision making based on the context about which behaviors should be activated and the selected behaviors are blended. In another work presented by, fuzzy behavior systems proposed for Autonomous navigation of Ground Vehicles in cluttered environment with unknown obstacles. Multivalue reactive fuzzy behaviors are used for arbitrating or fusing of the behaviors which action selection is relied on the available sensor information. In another work by Ramos et al., a hierarchical fuzzy decision-making algorithm introduced for behavior coordination of a robot based on arbitration mechanisms. In this method behaviors are not combined and just one behavior with maximum resulting value is selected and executed each time. A Fuzzy action selection approach was developed by Jaafar and McKenzie for navigation of a virtual agent. The fuzzy controller is comprised of three behaviors. The objective of this work is to solve the behavior's conflict. The method uses fuzzy α-levels to compute the behavior's weight and the Hurwicz criterion is used to select the final action. Wang and Liu introduced a new behavior-based navigation method called "minimum risk method". This behavior-based method applies the multi-behavior coordination strategy includes the global Goal seeking (GS) and the local Obstacle Avoidance (OA) (or boundary-following) behaviors. The fuzzy logic is applied to design and coordinate the proposed behaviors.

Fuzzy Control System in Mobile Robot Navigation

Design of a Fuzzy Controller

The schematic diagram of the fuzzy controller is shown in Figure. The fuzzy controller design steps include: 1) Initialization, 2) Fuzzification, 3) Inference and 4) Difuzzification.

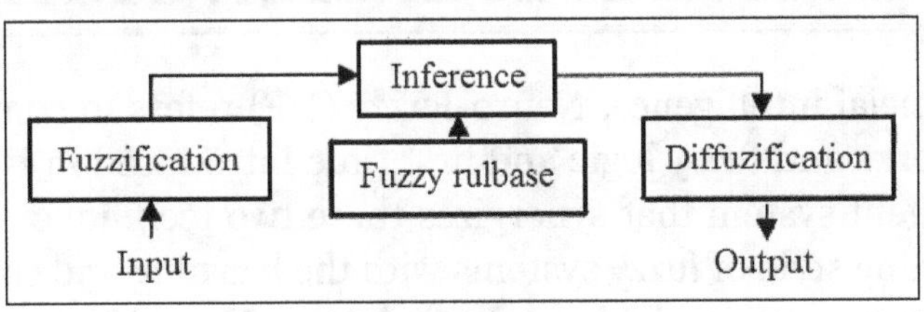

The fuzzy controller structure.

First step is identifying the linguistic input and output variables and definition of fuzzy sets (Initialization). Fuzzification or fuzzy classification is the process of converting a set of crisp data into a set of fuzzy variables using the membership functions (fuzzy sets). For example in Figure, the degree of membership for a given crisp is 0.6. Shape of the membership functions depends on the input data can be triangular, piecewise linear, Gaussian, trapezoidal or singleton.

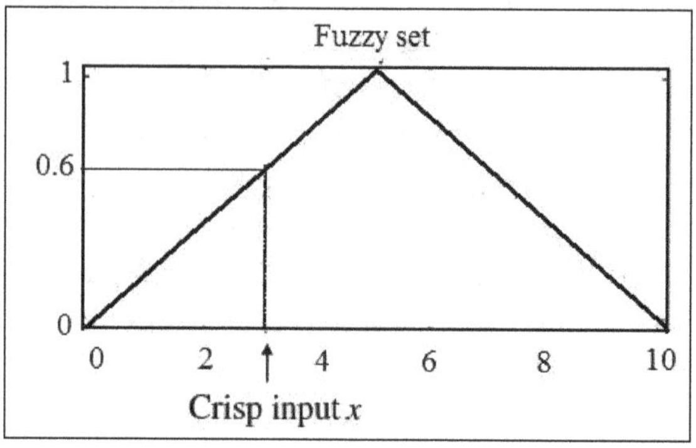

Membership degree of a crisp input x in the fuzzy set.

A rule base is obtained by a set of IF-THEN rules and inference evaluates the rules and combines the results of the rules. The final step is Defuzzification which is the process of converting fuzzy rules into a crisp output. An example of a simple fuzzy control system is shown in figure.

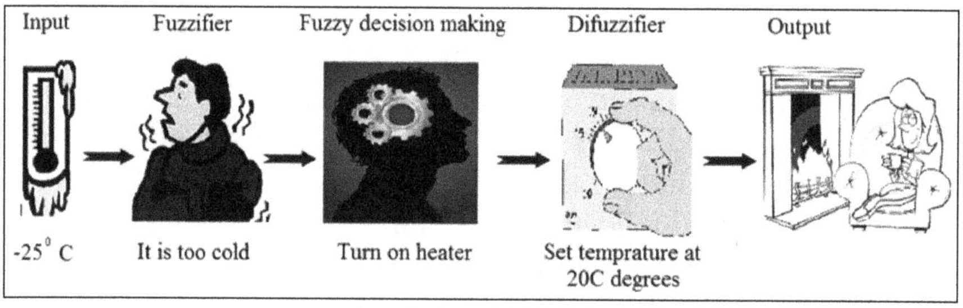

Example of a fuzzy control system.

INTELLIGENT NEURO-FUZZY APPLICATION IN SEMI-ACTIVE SUSPENSION SYSTEM

In the field of artificial intelligence, Neuro-Fuzzy (NF) refers to combinations of artificial neural networks and fuzzy logic and first time introduced in 1990s. Neuro-fuzzy results in a intelligent system that synergizes these two techniques by combining the human-like reasoning style of fuzzy systems with the learning and connectionist structure of neural networks. NF is widely termed as Fuzzy Neural Network (FNN) or Neuro-Fuzzy System (NFS) in the literature. NFS (the more popular term is used henceforth) incorporates the human-like reasoning style of fuzzy systems through the use of fuzzy sets and a linguistic model consisting of a set of IF-THEN fuzzy rules. The main strength of neuro-fuzzy systems is that they are universal approximations with the ability to solicit interpretable IF-THEN rules.

The strength of neuro-fuzzy systems involves two contradictory requirements in fuzzy modeling: interpretability versus accuracy. In practice, one of the two properties prevails. The neuro-fuzzy in fuzzy modeling research field is divided into two areas: linguistic fuzzy modeling that is focused on interpretability, mainly the Mamdani model; and precise fuzzy modeling that is focused on accuracy, mainly the Takagi-Sugeno-Kang (TSK) model.

The studies made full use of the advantages of the neural-network and the fuzzy logic controller and solved the different problems in suspension systems. Few researches involved combination of the two techniques to solve the time-delay and the inherent nonlinear nature of the Magneto-Rheological (MR) dampers in semi-active strategy for full car model with high degrees of freedom. For the intelligent system, fuzzy controller which inputs are relative velocities across MR dampers that are excited by road profile for predicting the force of MR damper to receive a desired passenger's displacement is applied. When predicting the displacement and velocity of MR dampers, a four-layer feed forward neural network, trained on-line under the Levenberg– Marquardt (LM) algorithm, is adopted. In order to verify the effectiveness of the proposed neuro-fuzzy control strategy, the uncontrolled system and the clipped optimal controlled suspension system are compared with the neuro-fuzzy controlled system. Through a numerical example under actual road profile excitation, it can be concluded that the control strategy is very important for semi-active control, the neuro-fuzzy control strategy can determine voltage of the MR damper quickly and accurately, and the control effect of the neuro-fuzzy control strategy is better than that of the other control strategies.

Full Car Model

In the full-car model, 11-DOFs is assumed, all wheels and passengers are dependent on each other and on the car's body. It is assumed that each wheel has an effect on the spring and damper of other wheels, and two axles of vehicle are dependent. MR actuator is utilized to damp the effect of road profile on the passengers. Note that MR shock absorber is added between the axel and car's body. In the full-car model, the effects of the rotations of the body around the roll and pitch axes are simulated. The suspension system using a full-car model has 11-DOFs, four of them for the four wheels, three for body displacement and its rotations and the last four for the passengers. Schematic of the full-car model with 11-DOFs and addition of the MR damper is shown in figure.

Where Mb, m_1, m_2, m_3, m_4, m_5, m_6, m_7 and m_8 stand for the mass of the car's body, mass of four wheels and mass of passengers, respectively. I_1 and I_2 are the moments of inertia of the car's body around two axes. The terms k_1, k_2, k_3, k_4, k_5, k_6, k_7 and k_8 are stiffness of the springs of the suspension system and stiffness of the springs of passengers seat, respectively. The terms k_{t1}, k_{t2}, k_{t3} and k_{t4} are stiffness of the tires. The terms b_1, b_2, b_3,

b_4, b_5, b_6, b_7 and b_8 are coefficients of car and passenger's seat dampers. Then, b_{r1}, b_{r2}, b_{r3} and b_{r4} are passive coefficients of the MR dampers, respectively. x_1, x_2, x_3, x_4, x_5, x_6, x_7, x_8, x_9, φ and θ indicate the DOFs of the suspension system model. The terms x_{i1}, x_{i2}, x_{i3} and x_{i4} indicate load profile disturbance, respectively. These parameters are used to clipped optimal strategy which is considered as a desire to train neural network and tuning fuzzy memberships. Here optimal force is depending on all state variables; therefore model with detail information is necessary.

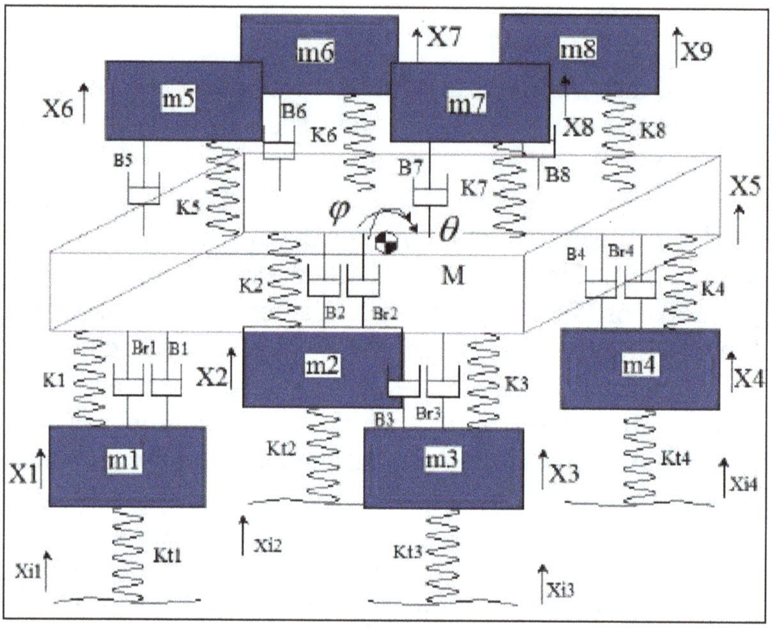

Full-car model with 11-DOFs.

Clipped Optimal Algorithm

The clipped optimal control strategy for an MR damper usually involves two steps. The first step is to assume an ideal actively–controlled device and construct an optimal controller for this active device. In the second step, a secondary controller finally determines the input voltage of the MR damper.

That is, the secondary controller clips the optimal force in a manner consistent with the dissipative nature of the device. The block diagram of the clipped optimal algorithm is shown in figure.

The clipped optimal control approach is to append a force feedback loop to induce the MR damper to produce approximately a desired control force f_c. The Linear Quadratic Regulator (LQR) algorithm has been employed both for active control and for semi-active control. Using this algorithm, the optimal control force f_c for f, which is force generated by an MR damper. Utilized clipped optimal algorithm for semi-active full car model.

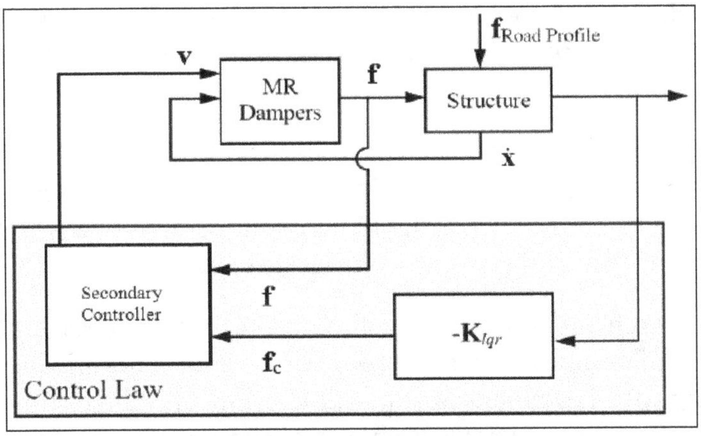

Clipped optimal algorithm block diagram.

Neuro-fuzzy Strategy using in Semi-active Vibration Control

Unfortunately, due to the inherent nonlinear nature of the MR damper to generate a force, a similar model for its inverse dynamics is difficult to obtain mathematically and also due to the nonlinearity of suspension system, its equations are complicated. Because of these reasons, a neural network with fuzzy logic controller is constructed to copy the inverse dynamics of the MR damper and suspension system.

Neuro-fuzzy controller is an artificial neural network, which is used to aggregate rules and to provide control result for the designed fuzzy logic controller. Application of fuzzy inference systems as a Fuzzy Logic Controller (FLC) has gradually been recognized as the most significant and fruitful application for fuzzy logic and fuzzy set theory. In the past few years, advances in microprocessors and hardware technologies have created an even more diversified application domain for fuzzy logic controllers, which range from consumer electronics to the automobile industry.

Indeed, for complex and/or ill-defined systems that are not easily subjected to conventional automatic control methods, FLCs provide a feasible alternative since they can capture the approximate, qualitative aspects of human reasoning and decision-making processes. However, without adaptive capability, the performance of FLCs relies exclusively on two factors: the availability of human experts, and the knowledge acquisition techniques to convert human expertise into appropriate fuzzy if-then rules and membership functions. These two factors substantially restrict the application domain of FLCs.

Consequently, a neural control design approach can usually be carried over directly to the design of fuzzy controllers, unless the design method depends directly on the specific architecture of the neural networks used. This portability endows us with a number of design methods for fuzzy controllers which can easily take advantage of a priori human information and expertise in the form of fuzzy if-then rules. The result of the above methodology is called Neuro-Fuzzy Control method. Neural and fuzzy logic controllers have been successfully implemented in the control of linear and nonlinear systems.

Unlike conventional controllers, such controllers do not require mathematical model and they can easily deal with the nonlinearities and uncertainties of the controlled systems. Also, a Levenberg-Marquardt (LM) neural controller has been designed for variable geometry suspension systems with MR actuators.

In the present research, an optimal controller Linear Quadratic Regulator (LQR) is designed for control of a semi-active suspension system for a full-model vehicle, using a neuro-fuzzy along with Levenberg-Marquardt learning and the results compared with Linear Quadratic Gaussian (LQG) (Zareh et al). The purpose in a vehicle suspension system is reduction of transmittance of vibrational effects from the road to the vehicle's passengers, hence providing ride comfort. To accomplish this, one can first design a LQR controller for the suspension system, using an optimal control method and use it to train a neuro-fuzzy controller. This controller can be trained using the LQR controller output error on an online manner.

Once trained, the LQR controller is automatically removed from the control loop and the neuro-fuzzy controller takes on. In case of a change in the parameters of the system under control or excitations, the LQR controller enters the control loop again and the neural network gets trained again for the new condition therefore it can ensure the robustness of strategy due to changes in excitations (Sadati et al). An important characteristic of the proposed controller is that no mathematical model is needed for the system components, such as the non-linear actuator, spring, or shock absorbers.

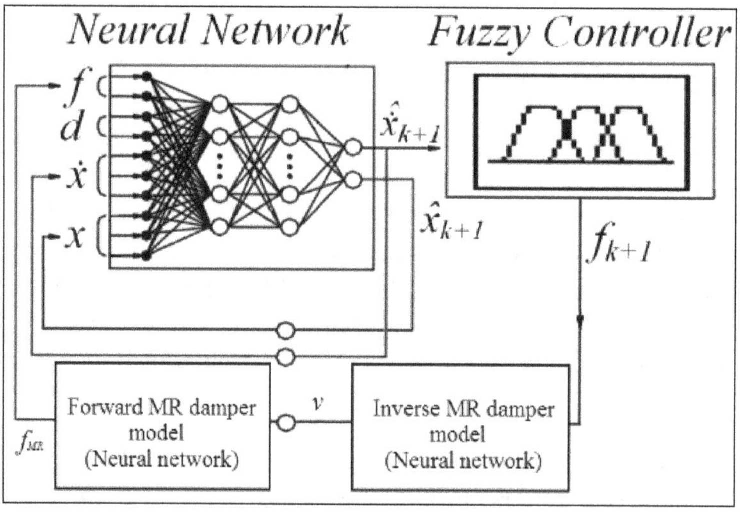

Architecture of the neuro-fuzzy control strategy.

The basic idea of the proposed neuro-fuzzy control strategy is that the forces of the MR dampers are determined by a fuzzy controller, whose inputs are the measured velocity response predicted by a neural network. The architecture of this strategy is shown in figure, which consists of two parts to perform different tasks. The first part is for the neural network to be trained on-line. The numbers of the sample data pairs are 3500, the training data pairs increase step by step during the entrance disturbance from road profile.

To select the network architecture, it is required to determine the numbers of inputs, outputs, hidden layers, and nodes in the hidden layers; this is usually done by trial and error. Therefore, one hidden layer, with six nodes, was adopted as one of the best suitable topologies for neural network.

The neural network is trained to generate the one step ahead prediction of the displacement $\hat{x}_{k+1}$ and the velocity $\hat{\dot{x}}_{k+1}$. Inputs to this network are the delayed outputs $\left(x_{k+3}, x_{k+2}, x_{k+1}, x_k, \dot{x}_{k+3}, \dot{x}_{k+2}, \dot{x}_{k+1}, \dot{x}_k\right)$, the delayed force which is predicted by fuzzy controller $\left(f_{k+1}\right)$, and the disturbance input $\left(d_k\right)$. At the initial time, the inputs of the network will be assigned the value of zero in accordance with the actual initial circumstance. Before online training, the network is trained off-line so as to obtain the weights that are as near to the desired value as possible.

The second part is the fuzzy controller, whose input is the measured relative velocity across MR dampers. The disturbance can be calculated by road profile model. The output of the fuzzy controller is the control force of the MR dampers. The main aim of this part is to determine the control force of the MR dampers quickly in accordance with the input excitation. In order to reach this aim, it is required to predict the responses of passengers in accordance with the optimal responses.

The third part is the feedforward neural network to be trained on-line to generate the required voltage of MR damper v. In fact, this part is the inverse dynamics model of MR damper.

This block diagram is designed by authors using of combination of advanced works. In this strategy there are three neural networks. First is to mapping of suspension system. Second is inverse model of MR damper and third is forward model of MR damper. The difference between inverse and forward model is their inputs and outputs where the inputs of inverse model is outputs of forward model and vice-versa. All data that are used to training, testing and validating are LQR results because, they are optimal and our desired.

Due to the inherent non-linear nature of the MR damper, a model for inverse dynamics of MR damper is difficult to obtain mathematically. Because of this reason, a feedforward back propagation neural network is constructed to copy the inverse dynamics of the MR damper. This neural network model is trained using input-output data generated analytically using the simulated MR model based on clipped algorithm. Using this inverse dynamics of MR damper, the required voltage signal v is calculated based on the desired control force f_c, the velocity of MR dampers $\hat{\dot{x}}_{k+1}$ &, and the displacement of MR damper x_{k+1}.

The fourth part is the feedforward back propagation neural network to be trained on-line in order to generate the MR damper forces fMR. The inputs of this neural network are voltage signal v, the velocity of MR damper $\hat{x}_{k+1}$ and the displacement of MR damper x_{k+1}. The difference between inverse and forward model is in inputs and outputs. The outputs of inverse model are the inputs of forward model.

The third and fourth part of the proposed neuro-fuzzy control strategy which is a three-layer feedforward neural network consists of an input layer with 3 nodes, a hidden layer with 6 nodes, and output layer with one node. Determining the numbers of inputs, outputs, hidden layers, and nodes in hidden layers of these three neural networks is done by trial and error. For all neural parts some of the corresponded results that are obtained by LQR are used as a desire data and some others are used as a testing data.

At the same time, the actual responses will feed back to the neural network and the weights and bias will be revised in real time. In this research, results from the optimal control history analysis method are used to simulate the actual measured responses. The errors between the predicted responses and the actual responses are used to update the weights of the neural network on-line.

The Neural Network based on Levenberg-Marquardt (LM) Algorithm

The MR damper forces based on the inputs of the reactive velocity. In such case, it is essential to develop an inverse dynamic model that predicts the corresponding control force which is to be generated by dampers.

Neural network is a simplified model of the biological structure which is found in human brains. This model consists of elementary processing units (also called neurons). It is the large amount of interconnections between these neurons and their capability to learn from data which makes neural network as a strong predicting and classification tool. a three-layer feed forward neural network, which consists of an input layer, one hidden layer, and an output layer, as shown in figure, is selected to predict the responses with MR dampers.

Here the networks are trained by LQR results (as a sample data). For example displacements, velocity and forces that are obtained by LQR are selected as a sample data for training and testing. Also target of networks are LQR results. For example in the second network (Inverse model of MR damper) the targets are voltages that obtained by LQR part of clipped method.

The net input value net_k of the neuron k in some layer and the output value Ok of the same neuron can be calculated by the following equations:

$$net_k = \Sigma\, w_{jk}\, O_j$$

$$O_k = f\left(net_k + \theta_k\right)$$

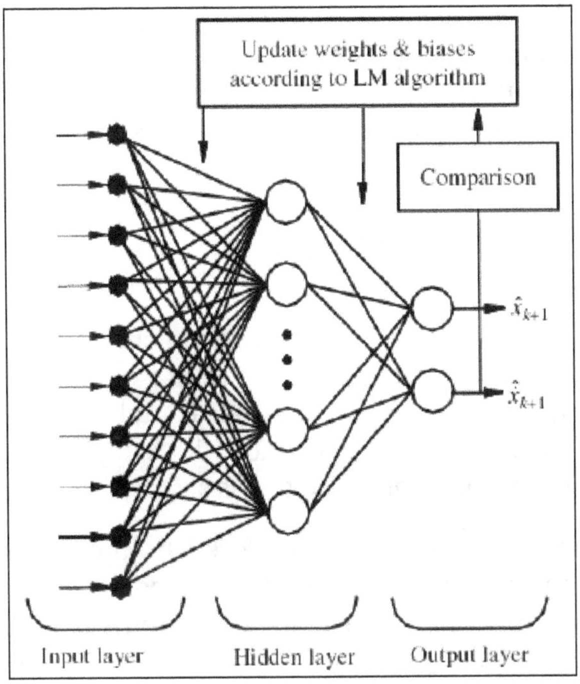

The neural network architecture.

where, w_{jk} is the weight between the jth neuron in the previous layer and the kth neuron in the current layer, O_j is the output of the jth neuron in the previous layer, f(.) is the neuron's activation function which can be a linear function, a radial basis function, and a sigmoid function, and y_k is the bias of the kth neuron. Feed forward neural network often has one or more hidden layers of sigmoid neurons followed by an output layer of linear neurons. Multiple layers of neurons with nonlinear transfer functions allow the network to learn nonlinear and linear relationships between input and output vectors. In the neural network architecture as shown in figure, the logarithmic sigmoid transfer function is chosen as the activation function of the hidden layer.

$$O_k = f\left(net_k + \theta_k\right) = 1/\left(1 + e^{-(net_k + \theta_k)}\right)$$

The linear transfer function is chosen as the activation function of the output layer.

$$O_k = f\left(net_k + \theta_k\right) = net_k + \theta_k$$

We note that neural network needs to be trained before it can predict any responses. As the inputs are applied to the neural network, the network outputs (.) are compared with the targets (.). The difference or error between both is processed back through the network to update the weights and biases of the neural network so that the network outputs match closer with the targets.

The input and output data are usually represented by vectors called training pairs. The process as is repeated for all the training pairs in the data set, until the network error converges to a threshold minimum defined by a corresponding performance function. In this research, the Mean Square Error (MSE) function is adopted (desired MSE is 1e-5). LM algorithm is adapted to train the neural network, which can be written as a following equation:

$$w^{i+1} = w^i - \left[\left(\delta^2 E / \delta w^{i2}\right) + \mu I\right]^{-1} \left(\delta E / \delta w^i\right)$$

where, i is the iteration index, $\delta E / \delta w^i$ is the gradient descent of the performance function E with respect to the parameter matrix w^i, μ ≥ 0 is the learning factor, and I is the unity matrix. During the vibration process, the neural network updates the weights and bias of neurons in real time in accordance with sampling pairs till the objective error is satisfied, i.e. the property of the system is acquired.

As we know, the main aim of the neural network is to predict the dynamic responses of the system, and to provide inputs to the fuzzy controller and also data for calculating the control force of MR dampers. Thus outputs of the neural network are predictions of displacement $\hat{x}_{k+1}$ and velocity $\hat{\dot{x}}_{k+1}$. In order to predict the dynamic responses of the system accurately, the most direct and important factors which affect the predicted dynamic responses are considered, i.e. the delayed outputs $\left(x_{k+3}, x_{k+2}, x_{k+1}, x_k, \dot{x}_{k+3}, \dot{x}_{k+2}, \dot{x}_{k+1}, \dot{x}_k\right)$, the predicted force $\left(f_{k+1}\right)$, and the disturbance input $\left(d_k\right)$. LM algorithm is encoded in Neural Networks Toolbox in MATLAB software.

Design of Fuzzy Controller

The first step of designing a fuzzy controller is determining the basic domains of inputs and outputs. The desired displacement and velocity responses are chosen as inputs of the fuzzy controller. The output of fuzzy controller is the control force of the MR damper, whose basic domain is -700N – 300N same as the working force of the MR damper calculated using LQR.

The membership functions are usually chosen in accordance with their characters and design experience.

For simplifying the calculation, triangular or trapezoidal functions are usually adopted as the membership functions. The triangular membership function is more sensitive to inputs than the trapezoidal form (Zh et al), in expectation that the control forces of the MR dampers are sensitive to excitations and responses, but in this case Gaussian and triangular forms are used because they have demonstrated better responses through trial and error. In this research, gaussian and triangular functions are adopted as the membership functions of velocity. The membership function curves of the velocity are shown in Figure. (Relative velocity across dampers).

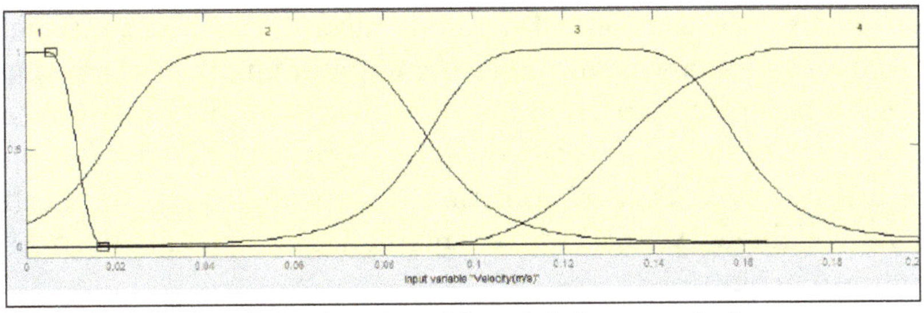

Membership function of front-left damper velocity.

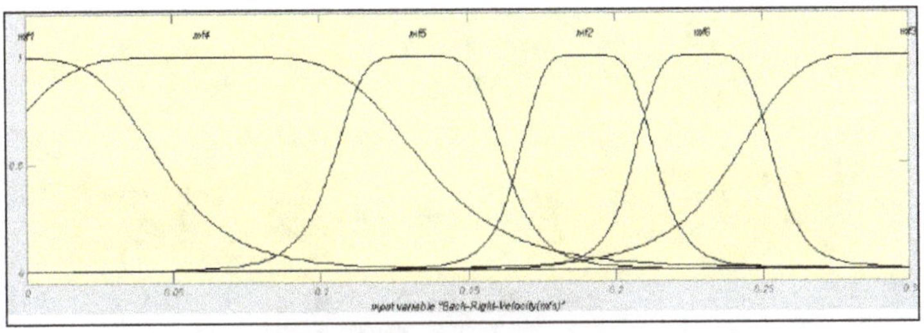

Membership function of front-right damper velocity.

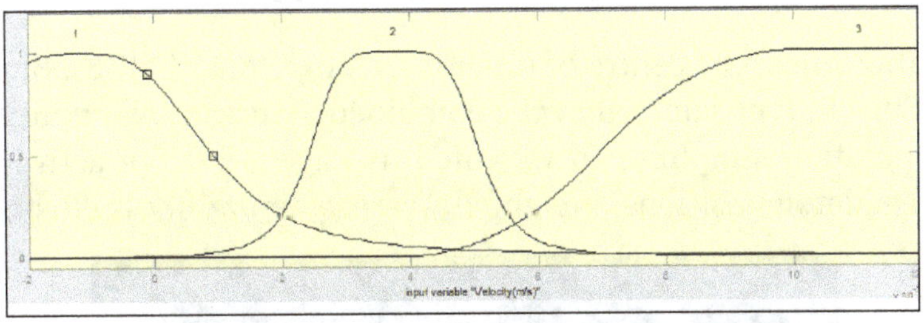

Membership function of back-left damper velocity.

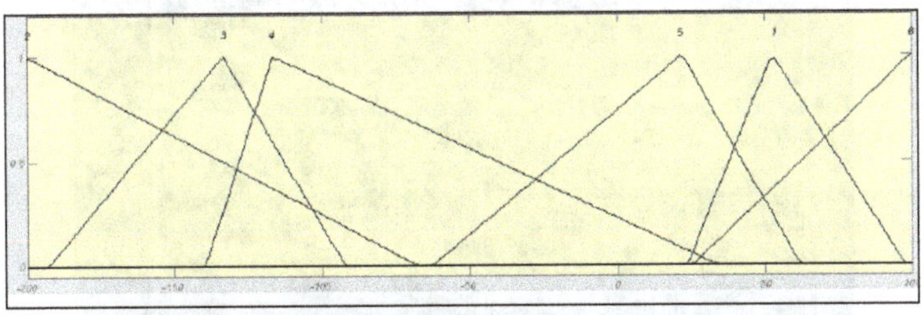

Membership function of back-right damper velocity.

Here, Sugeno inference engine with linear output is used, the main difference between Mamdani and Sugeno is that the Sugeno output membership functions are either linear or constant. It has led to reduction of computational cost because it does not need any defuzzification procedure. A Sugeno fuzzy model is computationally efficient platform that is well suited for implementation of non-linear associations through the construction of many piecewise linear relationships (Yen et al) .A typical rule in a Sugeno fuzzy model has the form:

If X is A1 and Y is B1 then Z = p1*x + q1*y + r1,

If X is A2 and Y is B2 then Z = p2*x + q2*y + r2,

where, q1 and q2 are constant. One of the main advantages of Sugeno method is well suited to mathematical analysis and is also computationally efficient, but Mamdani method is well suited to human input and it is intuitive. The basic idea of the fuzzy rules is that the control force increases with the increasing velocity responses. In this research, OR function is MAX, AND function is MIN and the defuzzification method is chosen as the Weighted Average (wtaver) method. The structure of considered fuzzy controller is shown in figure.

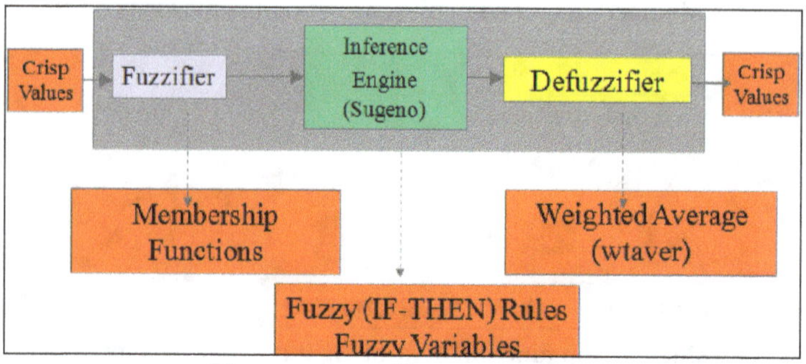

The structure of fuzzy controller.

For defuzzification we apply centre of gravity for singletons (COGS). Since we are implementing a Sugeno type controller, the combined activation, accumulation, and defuzzification operation simplifies to weighted average, with the activation strengths weighting the singleton positions. Weighted Average defuzzifier is illustrated in figure.

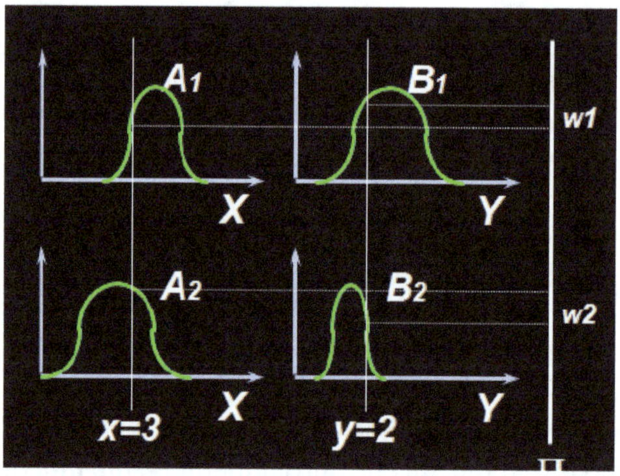

Sugeno-style rule evaluation.

z1 =p1*x+q1*y+r1

z2 =p2*x+q2*y+r2

Z=[w1*z1+w2*z2]/[w1+w2]

The membership function curves of the force for front-left damper as a fuzzy output (force vs. velocity) is shown in figure.

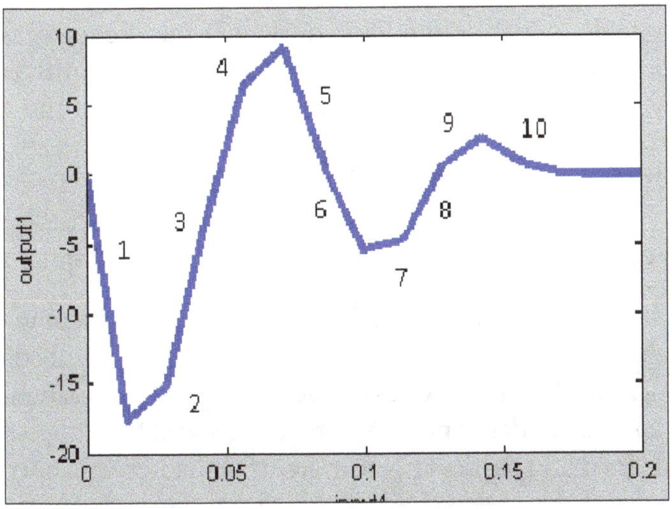

Membership function of back-right damper velocity
(force on vertical axis vs. velocity on horizontal axis).

The rule base used in the semi-active suspension system shown in table with fuzzy terms derived by the designer's knowledge and experience (because of shortage of space some of them are presented).

Table: Rule base.

Front-left	Front-right	Back-left	Back-right	Force
1	mf_3	1	2	1
1	mf_3	1	3	1
1	mf_4	2	3	6
1	mf_2	2	5	4
2	mf_6	1	5	6
2	mf_5	3	6	6
3	mf_6	2	1	8
3	mf_2	1	1	10
4	mf_5	3	1	1

APPLICATION OF FUZZY LOGIC IN CONTROL OF ELECTRICAL MACHINES

During the past decades, fuzzy logic control (FLC) has been one of the most active and fruitful areas for research in the application of fuzzy set theory.

The literature in fuzzy control has been growing rapidly in recent years, making it difficult to present a comprehensive survey of the wide variety of applications that have been made. Fuzzy logic, which is the logic on which fuzzy control is based, is much closer in spirit to human thinking and natural language than the traditional logical systems. Basically, it provides an effective means of capturing the approximate and the inexact nature of the real world. The fuzzy logic controller is a set of linguistic control rules related by the dual concepts of fuzzy implication and the compositional rule of inference. The FLC provides an algorithm which can convert the linguistic control strategy based on expert knowledge into an automatic control strategy.

The concept of FLC is to utilize the qualitative knowledge of a system to design a practical controller. For a process control system, a fuzzy control algorithm embeds the intuition and experience of an operator designer and researcher. The fuzzy control method is suitable for systems with non-specific models, and therefore, it suits well to a process where the model is unknown or ill-defined and particularly to systems with uncertain or complex dynamics.

The implementation of such control consists of translating the input variables to a language like: positive big, zero, negative small, etc. and to establish control rules so that the decision process can produce the appropriate outputs. Fuzzy control (FC) using linguistic information possesses several advantages such as robustness, model-free, universal approximation theorem and rules-based algorithm.

As an intelligent control technology, fuzzy logic control (FLC) provides a systematic method to incorporate human experience and implement nonlinear algorithms, characterized by a series of linguistic statements, into the controller. In general, a fuzzy control algorithm consists of a set of heuristic decision rules and can be regarded as an adaptive and nonmathematical control algorithm based on a linguistic process, in contrast to a conventional feedback control algorithm.

The fuzzy control also works as well for complex nonlinear multi-dimensional system, system with parameter variation problem or where the sensor signals are not precise. It is basically nonlinear and adaptive in nature, giving robust performance under parameter variation and load disturbance effect.

In process control applications, recent literature has explored the potentials of fuzzy control for machine drive application. It has been shown that a properly designed direct fuzzy controller can outperform conventional proportional integral derivative (PID) controllers.

Based on the analysis of the SM transient response and fuzzy logic, a fuzzy controller is developed. The fuzzy controller generates the variations of the reference current vector of the SM speed control based on the speed error and its change. Digital simulation results shows that the designed fuzzy speed controller realises a good dynamic behavior of the motor, a perfect speed tracking with no overshoot and a good rejection of impact loads disturbance. The results of applying the fuzzy logic controller to a SM show best performances and high robustness than those obtained by the application of a conventional controller (PI).

Fuzzy Logic Control

The structure of a complete fuzzy control system consists of the following main parts:

- Fuzzification,
- Knowledge base,
- Inference engine,
- Defuzzification.

Figure shows the internal configuration of a fuzzy logic controller.

Fuzzy Logic Principle

The fuzzification module converts the crisp values of the control inputs into fuzzy values. A fuzzy variable has values which are defined by linguistic variables (fuzzy sets or subsets) such as low, Medium, high, big, slow where each one is defined by a gradually varying membership function. In fuzzy set terminology, all the possible values that a variable can assume are named universe of discourse, and the fuzzy sets (characterized by membership functions) cover the whole universe of discourse. The shape of fuzzy sets can be triangular, trapezoïdale, etc.

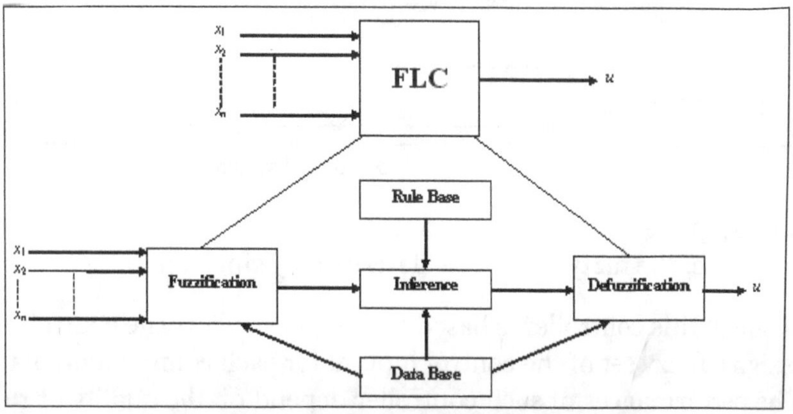

The internal configuration of a fuzzy logic controller.

A fuzzy control essentially embeds the intuition and experience of a human operator, and sometimes those of a designer and researcher. The data base and the rules form the knowledge base which is used to obtain the inference relation R. The data base contains a description of input and output variables using fuzzy sets. The rule base is essentially the control strategy of the system. It is usually obtained from expert knowledge or heuristics, it contains a collection of fuzzy conditional statements expressed as a set of IF-THEN rules, such as:

$$R^{(i)}: \text{If } x_1 \text{ is } F_1 \text{ and } x_2 \text{ is } F_2 \text{ .... and } x_n \text{ is } F_n \text{ THEN Y is } G^{(i)},\ i=1, ..., M$$

Where, $(x_1, x_2, ..., x_n)$ is the input variables vector, Y is the control variable, M is the number of rules, n is the number of fuzzy variables, $(F_1, F_2, ... F_n)$ are the fuzzy sets.

For the given rule base of a control system, the fuzzy controller determines the rule base to be fired for the specific input signal condition and then computes the effective control action (the output fuzzy variable).

The composition operation is the method by which such a control output can be generated using the rule base. Several composition methods, such as max-min or sup-min and max-dot have been proposed in the literature.

The mathematical procedure of converting fuzzy values into crisp values is known as 'defuzzification'. The choice of defuzzification methods usually depends on the application and the available processing power. This operation can be performed by several methods of which center of gravity (or centroïd) and height methods are common.

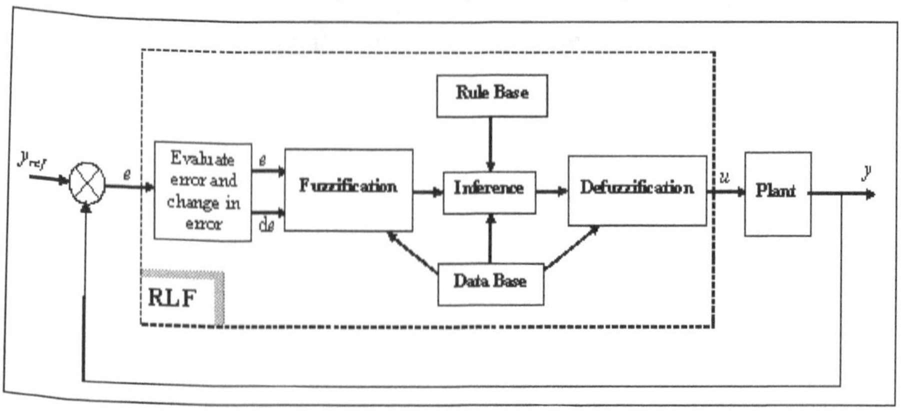

Basic structure of fuzzy control system.

The actual crisp input are approximates to the closer values of the respective universes of discourse. Hence, the fuzzy inputs are described by singleton fuzzy sets.

The elaboration of this controller is based on the phase plan. The control rules are designed to assign a fuzzy set of the control input u for each combination of fuzzy sets of e and Δe. The performances of such controller depend on the quality of rules and the choice of the fuzzy sets that describe number of the inputs and the output of the controller.

Fuzzy Control with Three Fuzzy Subsets

Table shows one of possible control rules based on three membership functions.

Table: Rules Base for speed control.

u	de		
	N	Z	P

e	N	N	N	Z
	Z	N	Z	P
	P	Z	P	P

The columns represent the rate of the error change de and the rows represent the error e. Each pair (*e, de*) determines the output level N to P corresponding to u.

Here N is negative, Z is zero, P is positive, are labels of fuzzy sets and their corresponding membership functions are depicted in figure. Figure shows the corresponding output surface.

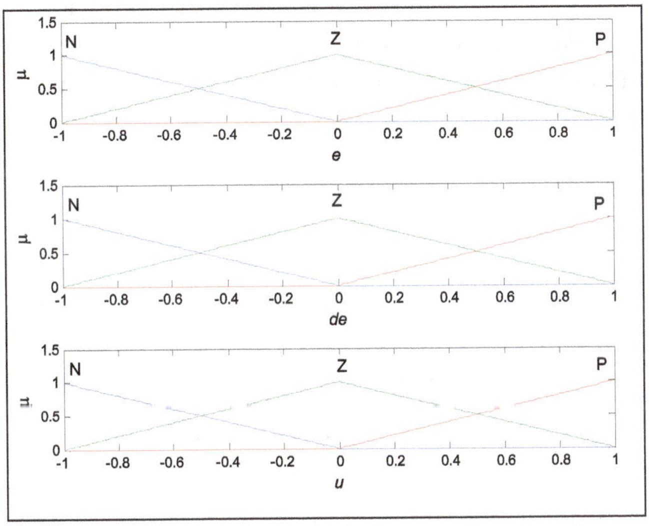

Membership functions for input e, de and u.

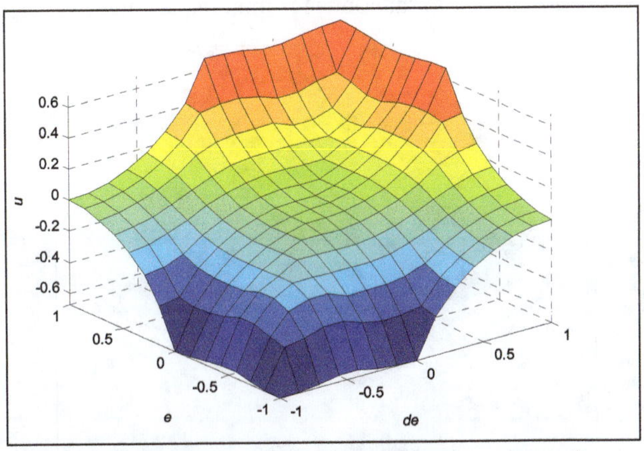

The output surface of the fuzzy inference system for three fuzzy subsets using the inputs and the output.

Fuzzy Control with Five Fuzzy Subsets

Table shows one of possible control rules based on five membership functions.

Table: Rules Base for speed control.

u		de				
		NB	NM	ZR	PM	PB
e	NB	NB	NB	NM	NM	ZR
	NM	NB	NM	NM	ZR	PM
	ZR	NM	NM	ZR	PM	PM
	PM	NM	ZR	PM	PM	GP
	PB	ZR	PM	PM	GP	GP

Here NB is negative big, NM is negative medium, ZR is zero, PM is positive medium and PB is positive big, are labels of fuzzy sets and their corresponding membership functions are depicted in figures. Figure shows the corresponding output surface.

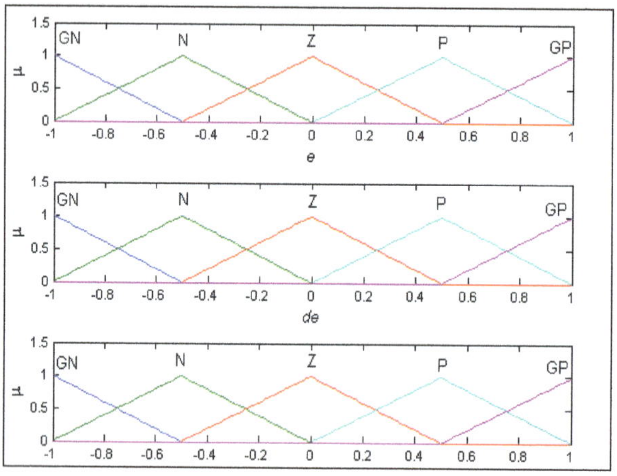

Membership functions for input e, de and u.

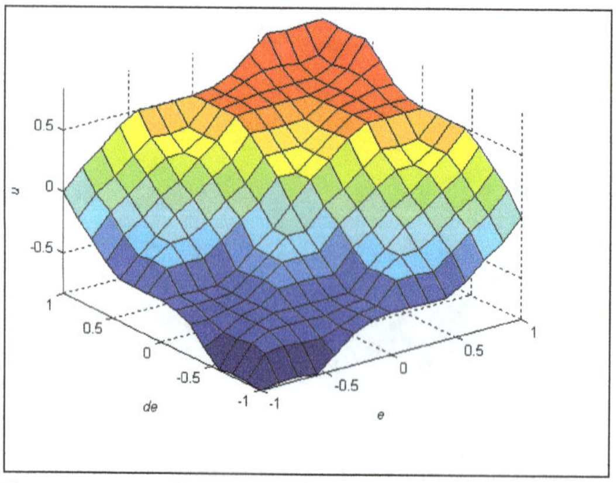

The output surface of the fuzzy inference system for five fuzzy subsets using the inputs and the output.

Fuzzy Control with Seven Fuzzy Subsets

Table shows one of possible control rules based on seven membership functions.

fuzzy PI is included in the controller. If the error is large and the controller tries to obtain a larger output value than the limits, the integral action will remain in pause until the correction level drops below the saturation level. So, as the error becomes smaller the integral action gains in importance as does the proportional action of the fuzzy PI controller. This second proportional action is used for fine tuning and to correct the response to sudden reference changes, helping to the proportional controller.

E_2, CE_2 and cu_2 are defined according to figure as,

$$E_2 = GE \bullet e,\ CE_2 = GCE \bullet ce,\ CU_2 = GCU \bullet cu_2$$

Where, GE , GCE and GCU are the scaling factors of the error, change of error and output, used to tuning the response of the controller. E_2 (error) and CE_2 (change of error) are the inputs of the fuzzy controller, an cu_2 (control action) is its output. Because the inputs of the fuzzy controller are the error and change of error it is useful to configure it as an incremental controller. This incremental controller adds a change to the current control signal of $\Delta U2_n$.

$$U2_n = U2_{n-1} + \Delta U2_n$$

And the $\Delta U2_n$. value in a PI controller would be,

$$\Delta U2_n = K_p \cdot \left(e_n - e_{n-1} + \frac{T_s}{T_i} e_n \right)$$

Where, Kp is the proportional gain and *Ts* and *Ti* the sample or control period and the integral time.

It is an advantage that the controller output $CU2_n$ is driven directly from an integrator, as it is then is easier to deal with windup and noise. The fuzzy PI controller output, $u2_n$, is called the change in output, and $U2_n$ is defined by,

$$U2_n = \sum_i (cu2_i \cdot GCU \cdot Ts)$$

The integrator will add only if $LIM_L < OUT_n < LIM_H$ and $cu2_i \neq 0$. The value of cu2 according to the inputs is,

$$cu2_n = f(GE \cdot e_n, GCE \cdot ce_n)$$

The function f is the fuzzy input-output map of the fuzzy controller. If it were possible to take the function f as a linear approximation, considering previous four equations, the gains related to the conventional PI would be,

$$K_p = GCE \cdot GCU$$

$$\frac{1}{T_i} = \frac{GE}{GCE}$$

These relations had shown the importance of the scaling factors. High values of GE produce a short rise time when a step reference is introduced but also a high overshot and a long settling time could arise. The system may become oscillatory and even unstable. If GE is low the overshot will decrease or disappear and the settling time increases. High values of GCE have the same effect as small values of GE and vice versa.

High values of GCU originate a short rise time and overshot when a step reference is introduced. If GCU is small the system gain is small and the rise time increases.

The global output value of the hybrid fuzzy controller is,

$$OUT_n = LIM_H \text{ if } U1_n + U2_n > LIM_H$$
$$OUT_n = LIM_L \text{ if } U1_n + U2_n > LIM_L$$

$$OUT_n = KP \cdot e_n + \sum_i^n \left(f\left(GE \cdot e_n, GCE \cdot ce_n\right) \cdot GCU \cdot Ts \right)$$

$$\text{if } LIM_L < U1_n + U2_n < LIM_H$$

The output of the controller is limited according to the maximum value of the hybrid fuzzy controller, for example for a speed controller the limit will be the maximum admissible torque and for the current controllers the limit will be the maximum admissible voltage of the machine.

For a practical implementation of the fuzzy controllers on a DSP the fuzzy membership functions of the antecedents and consequents are triangular and trapezoidal types because the calculus complexity is lower than the calculus complexity when are used Gaussian or Bell membership functions.

With the information of the plant model, the fuzzy sets and their linguistic variables are defined for the antecedents and consequents. The control strategy has to be implemented based on the engineer experience and if it is possible using simulation tools. The control strategy is stored in the rule-base in the form If-Then and an inference strategy will be chosen.

Then the system is ready to be tested to see if the closed-loop specifications are met. First simulations will be carried analyzing the dynamic behavior and the stability of the plant and finally the adjustment will be tested and adjusted again in the real machine control platform.

To get the rule-base of the controller the reference and feedback values are compared and the control action is determined to correct the deviation between reference and feedback. As an example, in the speed loop a positive increase of the speed error because the real speed is lower than the reference, must force to the controller to increase their output or torque reference, Te, to increase the machine speed as detailed in equation $T_e - T_L = J\dfrac{d\omega_m}{dt} + B\omega_m$. Something similar happens with the change of error; if the change of error is positive big, that means that the machine is decelerating, then the controller has to increase the torque to reduce the effect, so the controller has to produce a positive big output to increase the electromagnetic torque.

For another error and change of error combinations, the base-rule of table applied to the fuzzy controller shows a phase trajectory reducing the error as shown in figure. This is valid for the speed, flux and current loops. The base-rule of table characterizes the control objectives and it is shown as a matrix with the phase trajectory superimposed. The dynamic behavior of the controller to make zero the error will depend on the antecedents and consequents position, on the selected inference strategy, on the used defuzzification method and on the scaling factors.

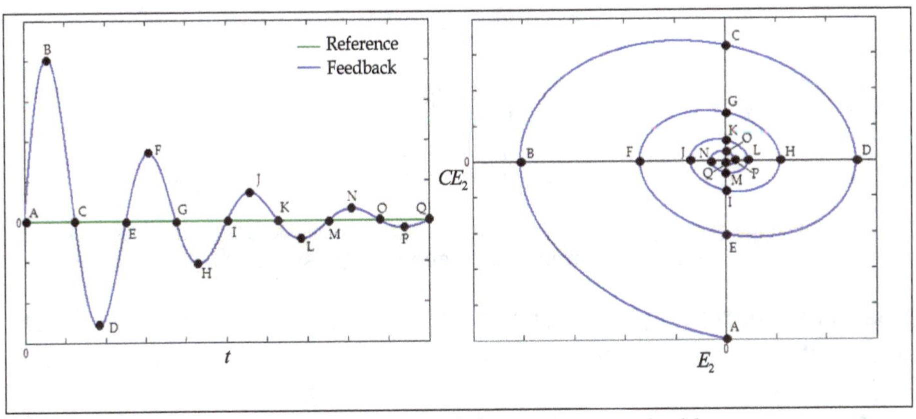

Fuzzy controller phase diagram when used table

The meaning of the linguistic terms used in table are: NB, negative big; NM, negative medium; NS, negative small; ZE, zero; PS, positive small; PM, positive medium and PB, positive big.

Table indicates the use of 49 rules. The first is read as,

If E_2 is Negative Big and CE_2 is Negative Big Then cu_2 is Negative Big

To adjust the scaling factors and the membership functions a first approximation is to make the controller as close as possible to a conventional PI controller.

The linguistic variable error and their linguistic terms position, figure, is the same for all fuzzy controllers. The error value is normalized for every controller, as an example when the speed error is 1000 rpm, their normalized value is 1.

Table: Rule-base of the fuzzy controller and phase diagram.

CE_2 \ E_2	NB	NM	NS	ZE	PS	PM	PB
NB	NB	NB	NB	NB	NM	NS	ZE
NM	NB	NB	NB	NM	NS	ZE	PS
NS	NB	NB	NM	NS	ZE	PS	PM
ZE	NB	NM	NS	ZE	PS	PM	PB
PS	NM	NS	ZE	PS	PM	PB	PB
PM	NS	ZE	PS	PM	PB	PB	PB
PB	ZE	PS	PM	PB	PB	PB	PB

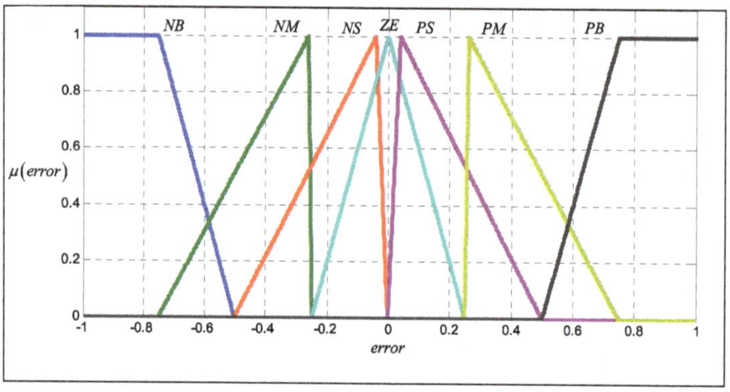

Linguistic variable error and its linguistic terms.

The linguistic variable change of error and their linguistic terms position, figure, is also the same for all fuzzy controllers. The change of error value is normalized for every controller.

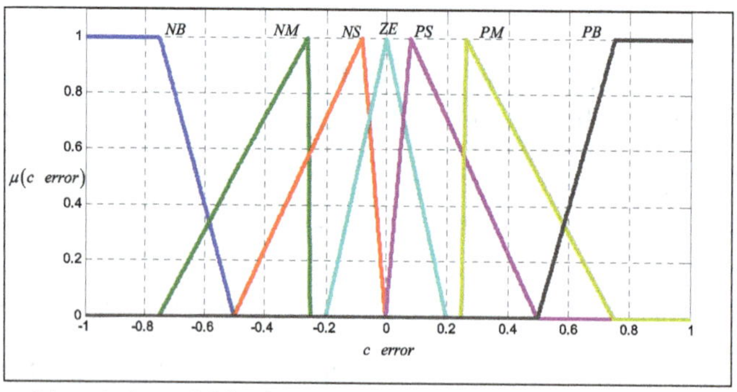

Linguistic variable change of error and its linguistic terms.

The linguistic variable of the control action or consequent and the position of its linguistic terms are shown in figure. The values are normalized, where a value of 20 in the real control action is normalized to 1.

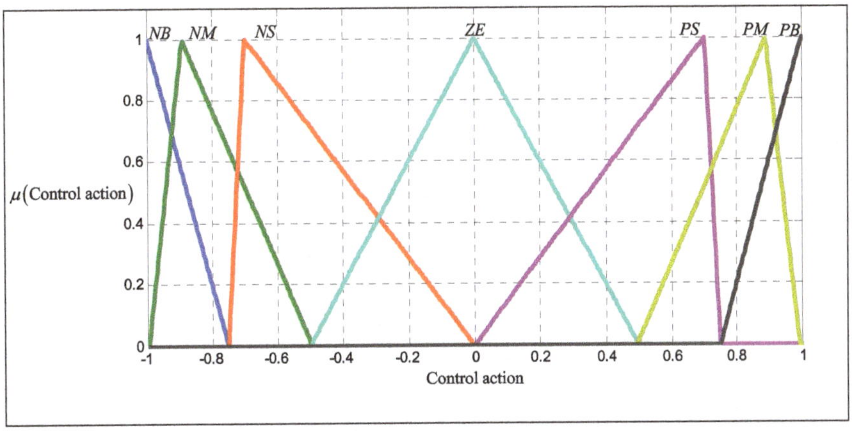

Control action linguistic terms.

In figure, the fuzzy controller surface can be seen. The used implication method is the AND method or min (minimum), which truncates the output fuzzy set and as aggregation the S-norm max (maximum) has been used. The used defuzzification method is the centroid or center of gravity, equation $y_o = \frac{\sum y\,\mu_T(y)}{\sum \mu_T(y)}$.

$$y_o = \frac{\sum y\,\mu_T(y)}{\sum \mu_T(y)}$$

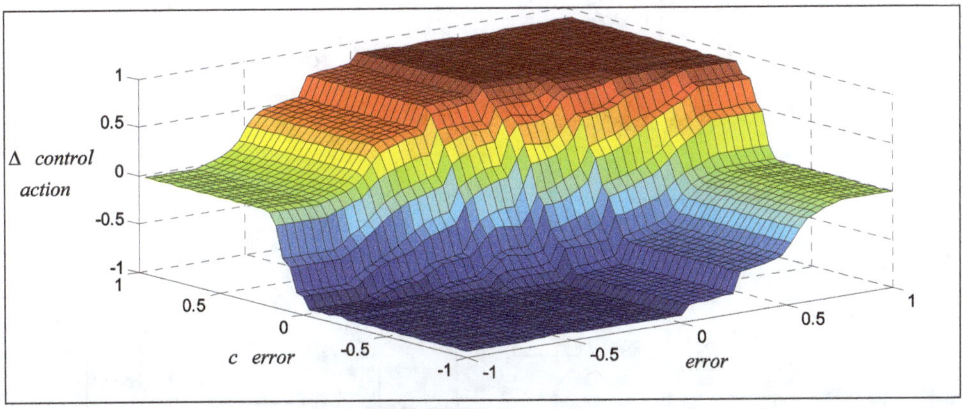

Fuzzy controller surface.

The linguistic variables are joined close to zero, showing a higher sensibility in this area. For this reason the slope of the surface in figure is high in a surrounding area around the point (0, 0, 0).

Squirrel-cage Machine Control

A schematic diagram of the induction motor indirect vector control with the fuzzy PI + P controllers is shown in figure. The scheme is obtained after operating with the machine equations and using the rotor flux reference system as shown in figure.

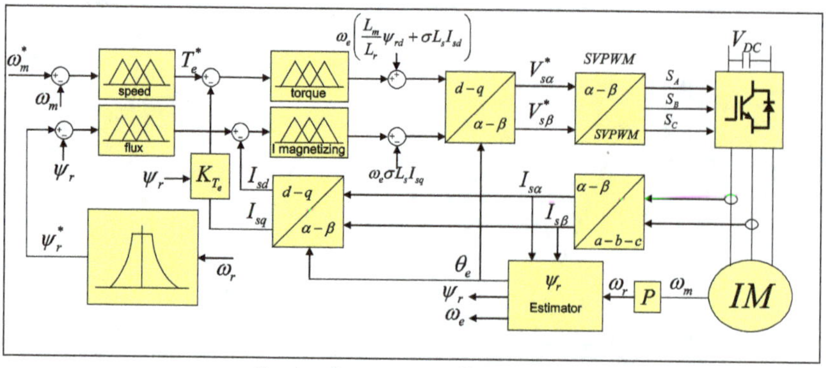

Squirrel cage control structure.

The rotor flux reference system makes possible the control of the AC machine as a DC machine, allowing the control of the machine torque with the stator current q component and the flux with the d component of the same current as can be deducted from equations $\bar{v}_{r,dq} = R_r \bar{i}_{r,dq} + \frac{d\bar{\psi}_{r,dq}}{dt} + j(\omega_e - \omega_r)\bar{\psi}_{r,dq}$ to $T_e - T_L = J\frac{d\omega_m}{dt} + B\omega_m$. A scheme showing these equations is shown in figure.

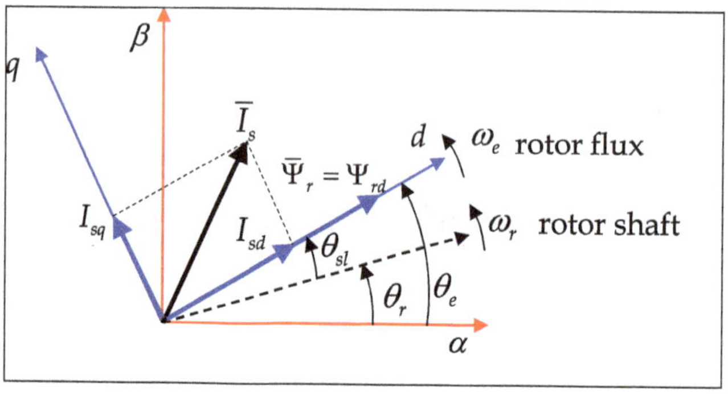

Fig: Rotor flux reference system

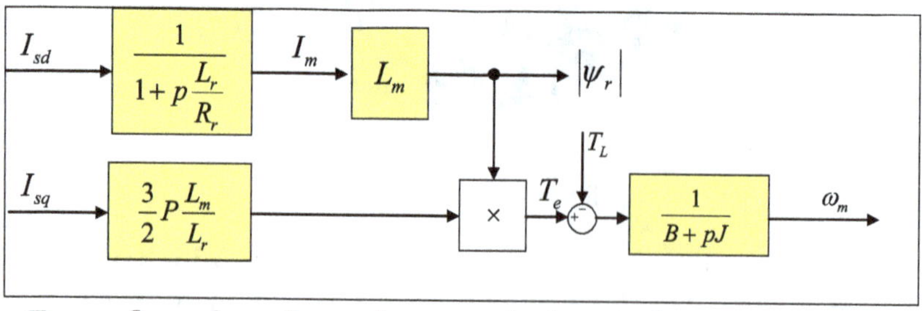
Torque, flux and speed control structure in the rotor flux reference system.

The speed error is the input of a hybrid fuzzy controller and the output of FPI+P controller will generate the torque producing stator current component command I_{sq}. The flux controller generates the flux producing stator current component I_{sd} according to the fluxspeed profile. Both currents are the input of two controllers to produce the stator voltages in the synchronous reference and then transformed to the stationary reference system to generate in the inverter the voltage vector for the motor.

The real platform to test the asynchronous motor and its main characteristics used also for the simulation purpose are shown in figure.

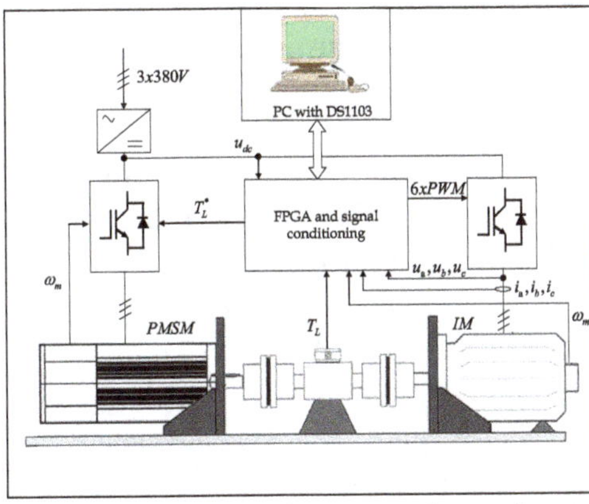

Voltage 380V III-Y	J = 0.038Kg*m^2
Frequency 50Hz	B = 0.008Kg*m^2/s P = 2 pole-pairs
Nominal current 14A	Rr = 0.57Ω Rs = 0.81Ω
Rated Torque 50Nm	Lm = 0.117774H Lr = 0.121498H
Rated speed 1440r.p.m.	Ls = 0.120416H

Induction motor rig test and asynchronous motor main characteristics.

The real system is based on a DS1103 board and is programmed using the software Matlab/Simulink. The board controls the IM inverter generating the SVPWM pulses. The speed is measured with a 4096 impulse encoder via a FPGA connected to the DS1103 using the multiple period method.

Torque or Current Control

As mentioned and shown in figure, the torque of the machine is controlled with the stator current q component and the flux with the d component. The relation between the torque Te and the stator current q component is,

$$T_e = \underbrace{\frac{3}{2} P \frac{L_m}{L_r} \psi_r}_{K_{T_o}} I_{sq}$$

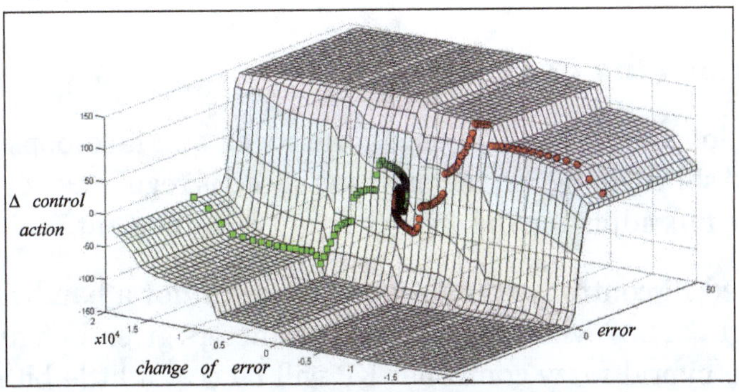

Stator current q component controller fuzzy surface and trajectory after current step of figure.

So first, torque and current magnetizing controllers will be adjusted. In a classical PI controller the proportional term for a bandwidth of 2500 r_{ad}/s and a phase margin of

80° with the machine parameters given in figure is 0.05. For the adjustment of the hybrid fuzzy controller KP will be 0.025, half of the proportional term in the PI. The scaling factors adjusted after simulations for the current controllers are GE = 150, GCE = 0.03 and GCU = 8. The regulators maximum and minimum limits are ±310V, the maximum motor phase voltage.

Figure shows the hybrid fuzzy stator q current controller surface and the trajectory when a step reference of -20 amperes is produced, and after 200 ms another step of 20 amperes as shown in figure is applied to the torque controller.

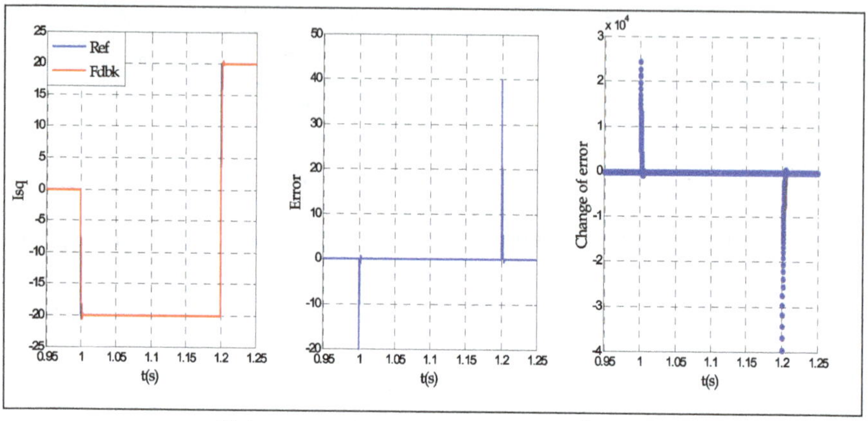

Stator current q component step reference
and feedback, error for the step, and change of error.

When the step reference is -20 amperes the feedback or real stator q current reaches the real value quickly, it takes 2 ms. The trajectory on the fuzzy surface for this step is the green line in the surface showing how the change of error and the error are decreasing to zero in about 2 ms. When the step reference goes from -20 to 20 amperes the feedback or real stator q current reaches the real value in 3 ms. The trajectory on the fuzzy surface for this step is the red line in the surface showing how the change of error and the error are decreasing to zero due to the value of the control action.

Speed and Rotor Flux Control

Once the current loops have been adjusted, the speed and flux loops will be adjusted. As mentioned and shown in figure, the machine speed is regulated adjusting the torque command and the flux adjusting the stator current d component.

In a classical speed PI controller the proportional term for a bandwidth of 750 rad/s and a phase margin of 80° with the machine parameters given in figure is 0.5. For the adjustment of the hybrid fuzzy controller KP will be 0.4, a little bit smaller than the proportional term in the PI. The scaling factors adjusted after simulations for the speed controllers are GE = 2, GCE = 0.01 and GCU = 300. The regulators maximum and minimum limits are ±50 Nm, the maximum motor torque or a stator current q component of 20 amperes.

Figure shows the hybrid fuzzy speed controller surface and the trajectory when a step reference from -1000 rpm to 1000 rpm and again to -1000 rpm as shown in figure is applied to the speed controller.

When the step goes from -1000 to 1000 rpm the trajectory on the fuzzy surface for this step is the green line, showing how the change of error and the error are decreasing to zero in about 180 ms. When the step reference goes from 1000 to -1000 rpm the feedback or real speed reaches the real value in 180 ms. The trajectory on the fuzzy surface for this step is the red line, showing how the change of error and the error are decreasing to zero due to the value of the control action.

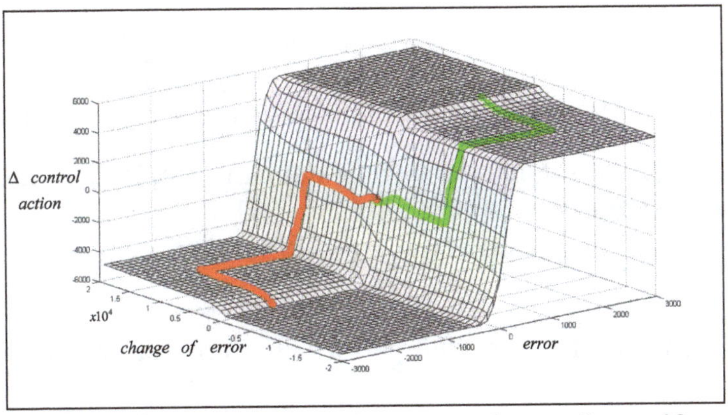

Speed controller fuzzy surface and trajectory after speed step of figure.

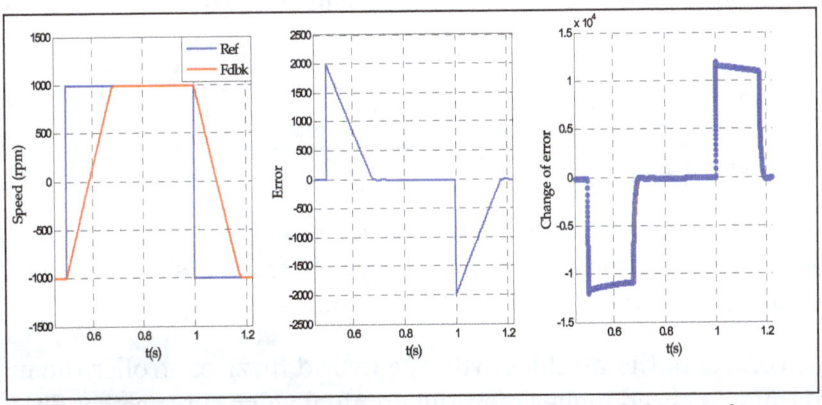

Speed step reference and feedback, error for the step, and change of error.

When the change of error is high, the controller output is at its maximum limit, and when the change of error decreases the control action also decreases close to zero as it can be seen in the trajectory of figure. The error and change of error trajectory of the surface in figure correspond to the values represented in figure. The control action contribution can be obtained from the fuzzy controller surface.

Figure shows the response of the real asynchronous motor of figure when a speed step is applied to the machine and later a load torque of 40 Nm after 0.3 s. Three classes of speed controllers are tested to see the response and compare them. A classical PI controller with a 750 rad/s and a phase margin of 80°, the adjusted hybrid Fuzzy PI + P controller and a Fuzzy controller without the KP term and GE = 2, GCE = 0.06 and GCU = 300 .

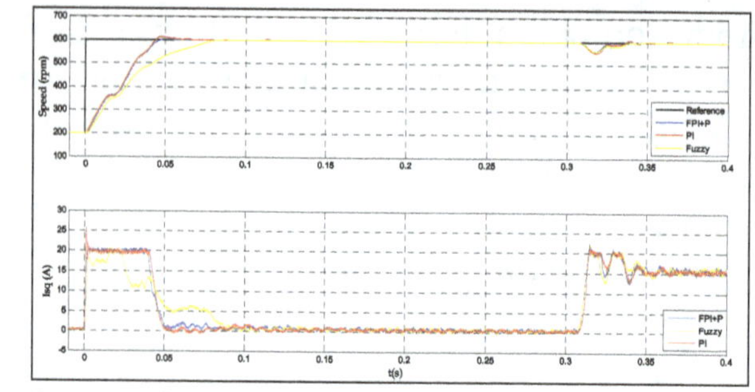

Top, speed step and response when PI, Fuzzy and Fuzzy PI + P controllers are used. Bottom, torque current controllers output

To compare the controllers, table shows time domain specifications and performance criteria, integrated absolute error (IAE), the integral of time-weighted absolute error (ITAE), the integral of the square of the error, ISE, and the integral of time multiply squared error (ITSE).

Table: Time domain specifications and performance criteria for three classes of controllers.

	Delay Time	Rise Time	Settling Time	% Overshoot	IAE	ITAE	ISE	ITSE
PI	1.4ms	42ms	56ms	3	97470	6754	2.23e7	3.29e5
Fuzzy	3.2ms	77ms	80ms	0	1.28e5	7579	2.86e7	4.8e5
FPI+P	1.4ms	42ms	47ms	0	96270	6000	2.23e7	3.01e5

Very similar results are obtained with the PI and FPI+P controllers, although according to the performance criteria the hybrid fuzzy controller is slightly better. The worst controller is the fuzzy controller as it is shown in table and figure.

To check the control of the machine with the hybrid fuzzy controller the machine will be forced to run at a speed higher than the nominal value. In such conditions the machine rotor flux has to decrease because the inverter DC voltage can't be higher, so the torque and stator current q component relation is changing as shown in equation $T_e = \underbrace{\frac{3}{2} P \frac{L_m}{L_r} \psi_r}_{K_{T_0}} I_{sq}$ and figure. This change should be taken in consideration in a classical PI regulator. In the hybrid fuzzy controller the adjustment done with the linguistic variables and the scaling factors shows that the control works properly. In figure, the left signals correspond to the real signals obtained whit the machine of the test rig and the right side signals are the simulated in the same conditions than the real case.

Because the speed is higher than nominal value, the flux decreases below the nominal value, to do this the stator current d component decreases and increases when the flux is increasing to the nominal value. The q component of the stator current related with the torque increases when the machine is accelerating and decreases when the machine decelerates.

The speed regulation in the flux weakening region is good, and real platform signals and simulations corroborate the hybrid fuzzy good performance.

The flux hybrid fuzzy controller scaling factors are GE = 200, GCE = 20 and GCU = 100. To evaluate the flux regulation, the rotor flux reference and feedback values could be compared in the flux weakening shown in figure. Both are very similar showing a very good flux regulation and the flux controller output corresponds with the stator current d component shown in the same figure.

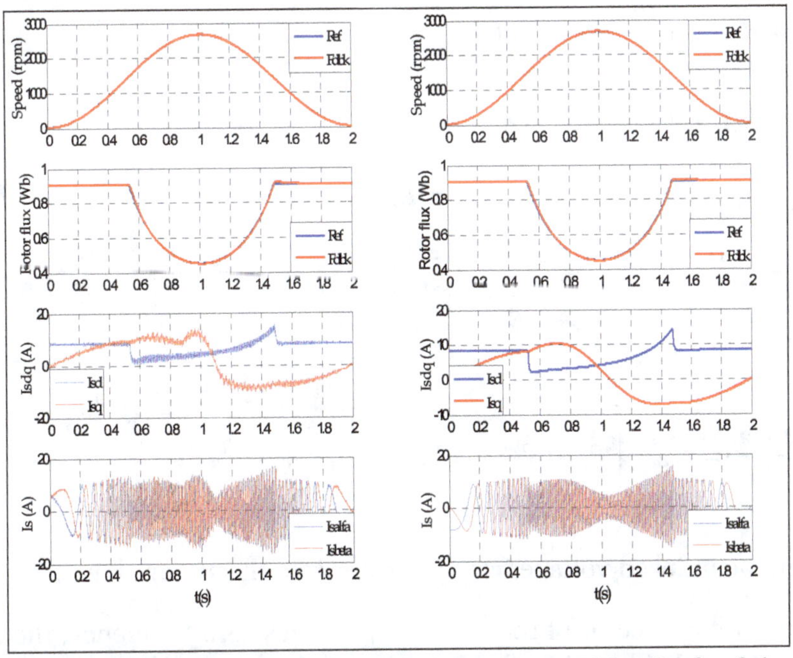

Left, real machine signals, speed, flux and stator currents. Right, simulated signals.

Speed Estimation

There are in literature many techniques of sensorless control. The first group is based on the fundamental mathematical model of the machine, that is, the flux density distribution in the air gap is sinusoidal. All these models depend on the machine parameters so the accuracy of the estimators will depend on different manner of the precision of these parameters. It is not possible with these techniques to achieve a stable and precise operation at very low speed.

The second group of techniques is based on the anisotropic properties of the machine.

Techniques like rotor slot ripple or main inductance saturation are used in this group.

From equations:

$\bar{v}_{r,dq} = R_r \bar{i}_{r,dq} + \frac{d\psi_{r,dq}}{dt} + j(\omega_e - \omega_r)\bar{\psi}_{r,dq}$ and $\bar{\psi}_{r,dq} = L_r \bar{i}_{r,dq} + L_m \bar{i}_{r,dq}$ and $L_r = L_m + L_{lr}$

considering rotor voltage zero, and after Laplace transformation of the respective space vectors the rotor flux will be,

$$\bar{\psi}_{r,dq}(p) = \frac{L_m}{1 + \frac{L_r}{R_r} p + j(\omega_e - \omega_r)\frac{L_r}{R_r}} \bar{i}_{s,dq}(p)$$

Operating with equations:

$$\bar{v}_{s,dq} = R_s \bar{i}_{s,dq} + \frac{d\psi_{s,dq}}{dt} + j\omega_e \bar{\psi}_{s,dq}$$

$$\bar{v}_{r,dq} = R_r \bar{i}_{r,dq} + \frac{d\bar{\psi}_{r,dq}}{dt} + j(\omega_e - \omega_r)\bar{\psi}_{r,dq}$$

$$\bar{\psi}_{s,dq} = L_s \bar{i}_{s,dq} + L_m \bar{i}_{r,dq} \quad \text{and} \quad L_s = L_m + L_{ls}$$

$$\bar{\psi}_{r,dq} = L_r \bar{i}_{r,dq} + L_m \bar{i}_{r,dq} \quad \text{and} \quad L_r = L_m + L_{lr}$$

The next equation is obtained,

$$\sigma L_s \frac{d\bar{i}_{s,dq}}{dt} = \bar{v}_{s,dq} - \left[R_s + \left(\frac{L_m}{L_r}\right)^2 R_r + j\omega_e\right]\bar{i}_{s,dq} + \underbrace{\frac{L_m}{L_r}\left(\frac{R_r}{L_r} - j\omega_r\right)\bar{\psi}_{r,dq}}_{\bar{vr}_{s,dq}}$$

It can be seen the induced voltage from the rotor into the stator as $\bar{v}r_{s,dq}$.

As the feeding voltage vector of the stator approaches zero frequency, the rotor speed approaches zero. If the equation above is observed in the stationary reference frame, $\omega_e = 0$, and using equation:

$$\bar{\psi}_{r,dq}(p) = \frac{L_m}{1 + \frac{L_r}{R_r} p + j(\omega_e - \omega_r)\frac{L_r}{R_r}} \bar{i}_{s,dq}(p) \text{ and } \bar{v}r_{s,dq}$$

It is calculated when $p \to 0$,

$$\bar{vr}_{s,\alpha\beta} \Big|_{\omega_{r \to 0}} = \lim_{p \to 0} \bar{vr}_{s,\alpha\beta} = \frac{L_m^2 R_r}{L_r^2} \bar{i}_{s,\alpha\beta}$$

The equation $\overline{vr}_{s,\alpha\beta} \mid \omega_{r\to 0} = \lim_{p\to 0} \overline{vr}_{s,\alpha\beta} = \frac{L_m^2 R_r}{L_r^2} \bar{i}_{s,\alpha\beta}$ is independent of ω_r when stator frequency is close to zero, so the variations of rotor speed have no influence on the stator equation 18 and this makes impossible to detect a speed variation on the stator current. So the mechanical speed of the rotor becomes not observable. Instead of this, when the magnitude of the induced voltage from the rotor into the stator is substantial, its value can be determined and the rotor state variables are then observable. So, there will be a limitation for very low speed operation due to the dc offset components in the measured stator currents and voltages.

The minimum stator frequency must be superior to zero to have an appropriate relation between induced voltage from the rotor into the stator and also to reduce the noise and parameters mismatch influence.

The rotor speed estimator used, figure, is based on the fundamental mathematical model of the machine. The rotor speed is obtained with the derivative of the rotor flux angle minus the slip speed,. The precision of the estimator has a great dependence on motor parameters and at low speeds a small error (offset for example) in the stator voltage can suppose an estimation error.

The rotor flux estimator contains two models, the open loop current model, which is supposed to produce an accurate estimation at low speed range, and an adaptive voltage model for a medium high speed range of operation. The transition between both models is adjusted by two hybrid fuzzy controllers, reducing the problems due to stator resistance and pure integrators at low speed.

The stator flux in the fixed reference frame related to the rotor flux and the stator current is,

$$\overline{\psi}^{i}_{s,\alpha\beta} = \frac{L_m}{L_r}\overline{\psi}^{i}_{r,\alpha\beta} + \frac{L_s L_r - L_m^2}{L_r}\bar{i}_{s,\alpha\beta}$$

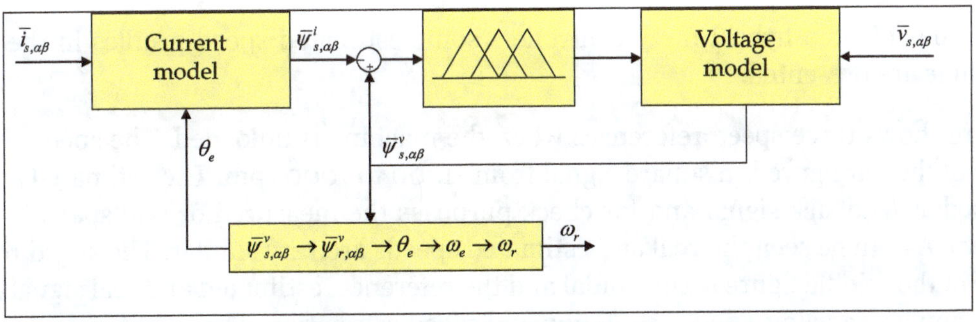

Rotor speed estimation using hybrid fuzzy controllers.

The stator flux using the voltage model is corrected by a compensation term, generated by two hybrid fuzzy controllers,

$$\overline{\psi}^{v}_{s,\alpha\beta} = \int\left(\overline{v}_{s,\alpha\beta} - R_s\overline{i}_{s,\alpha\beta} - \overline{v}_{comp}\right)$$

And

$$\overline{v}_{comp_n} = KP\cdot\left(\overline{\psi}^{v}_{s,\alpha\beta} - \overline{\psi}^{i}_{s,\alpha\beta}\right) + \sum_{i}^{n}\left(f\left(GE\cdot\left(\overline{\psi}^{v}_{s,\alpha\beta} - \overline{\psi}^{i}_{s,\alpha\beta}\right)_n,\right.\right.$$

$$\left.\left.GCE\cdot c\left(\overline{\psi}^{v}_{s,\alpha\beta} - \overline{\psi}^{i}_{s,\alpha\beta}\right)_n\right), GCU\cdot Ts\right)$$

With the obtained stator flux, the rotor flux and angle according to the voltage model are determined,

$$\overline{\psi}^{v}_{r,\alpha\beta} = \frac{L_r}{L_m}\overline{\psi}^{v}_{s,\alpha\beta} + \frac{L_s L_r - L^2_m}{L_m}\overline{i}_{s,\alpha\beta}$$

And

$$\theta_e = \theta_{\psi r} = \tan^{-1}\frac{\psi^{v}_{r\beta}}{\psi^{v}_{r\alpha}}$$

Finally the rotor speed is obtained,

$$\omega_r = \omega_{\psi r} - \omega_{sl} = \frac{d\theta_{\psi r}}{dt} - \frac{L_m R_r}{L_r\left(\psi^2_{r\alpha} + \psi^2_{r\beta}\right)}\left(\psi_{r\alpha} i_{s\beta} + \psi_{r\beta} i_{s\alpha}\right)$$

The scaling factors adjusted after simulations for the hybrid fuzzy controllers are, KP = 245, GE = 105, GCE = 1 and GCU = 11.

With the adjusted hybrid fuzzy controllers some estimated speed profiles in the real machine are presented.

Figure shows three speed references when the machine is unloaded. The speed reference of the left figure is a square signal from -1000 to 1000 rpm. The estimated speed is used as feedback signal and for check purposes the measured or real speed is also shown. As can be seen the real and estimated speeds are very similar. The speed reference of the middle figure is sinusoidal and the reference, estimated and real signals are very similar, showing a good regulation and speed estimation. The right figure shows a random speed reference crossing during 2 seconds at a speed close to zero rpm, where the speed is poorly observable. The reference, estimated and real signals are very similar even at zero speed for a short time.

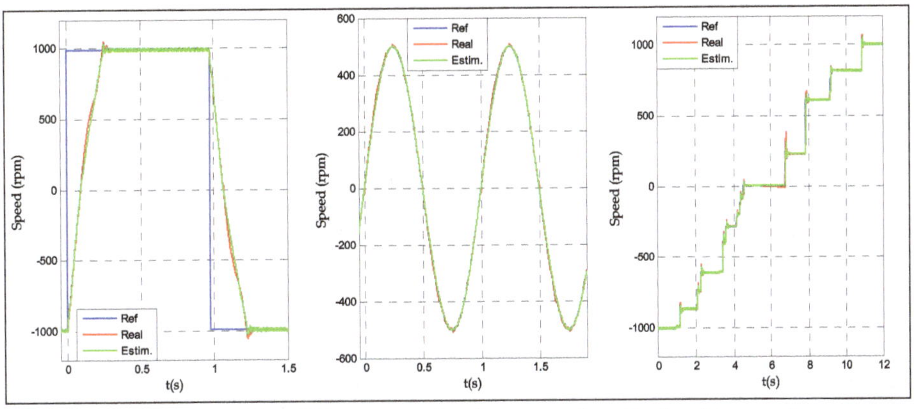

Sensorless control for different speed references when the load torque is cero.

Figure shows the speed estimation when a load perturbation of ±30 Nm is applied to the machine. There is an error between the real speed and the estimated speed when the machine is loaded due to parameters mismatch.

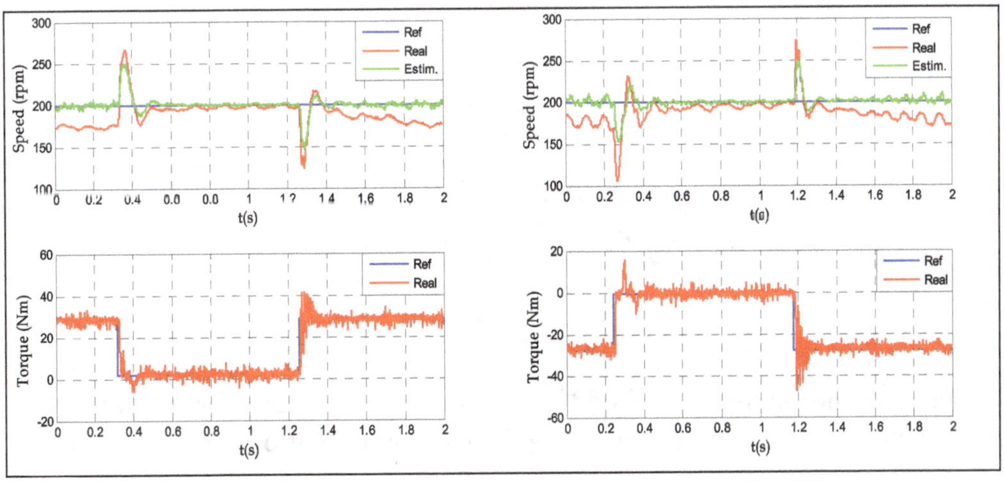

Sensorless control for 200 rpm and torque step loads of ±30Nm.

Doubly Fed Induction Generator Control

A doubly fed induction generator (DFIG) vector control with the fuzzy PI + P controllers is shown in figure. The scheme is obtained after operating with the machine equations and using the stator flux reference system shown in figure.

The converter Back to Back configuration provides to the DFIG the ability of reactive power control. Using the appropriate reference system it is possible to decouple the active and reactive power control by the independent control of the rotor excitation current. Due to the bi-directional power converter in the rotor side, the DFIG is able to work as a generator in the sub-synchronous (slip speed is positive, $s > 0$) and super-synchronous (slip speed is negative, $s < 0$) operating area.

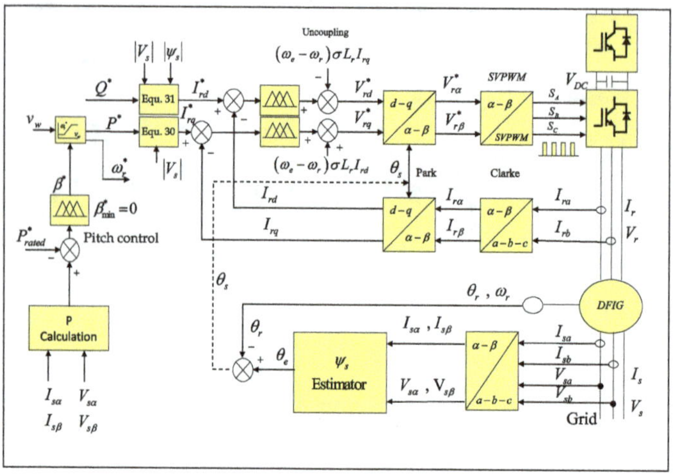

DFIG control structure.

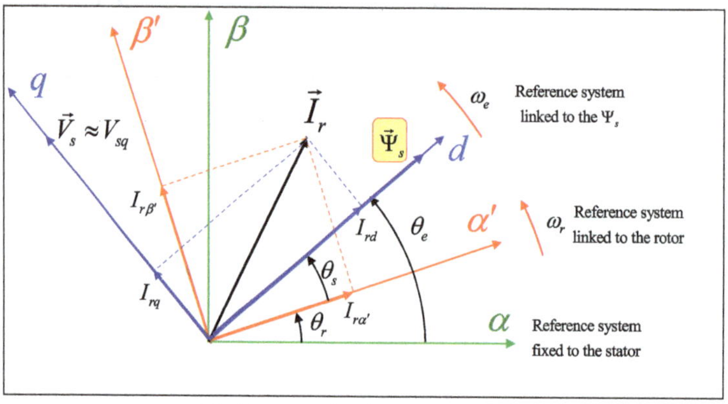

DFIG control reference systems.

When the reference system is linked to the stator flux, the stator flux q component is zero, and when operating with equation 3 the next two equations are obtained,

$$i_{sd} = \frac{|\overline{\psi}_s|}{L_s} - \frac{L_m}{L_s} i_{rd}$$

$$i_{sq} = -\frac{L_m}{L_s} i_{rq}$$

This means that the stator current can be controlled with the rotor current. Taking into account that the stator resistance is small, the stator flux can be considered constant and its value is,

$$|\overline{\psi}_s| \approx \frac{|\overline{V}_s|}{\omega_e}$$

The stator voltage d component is almost zero because the reference system is oriented along the stator flux, so considering that the stator active and reactive power is,

$$P_s = \frac{3}{2}\left(v_{sd}i_{sd} + v_{sq}i_{sq}\right) \text{ and } Q_s = \frac{3}{2}\left(v_{sq}i_{sd} - v_{sd}i_{sd}\right)$$

It can be obtained that,

$$P_s \approx -\frac{3}{2}\omega_e \Psi_s \frac{L_m}{L_s} i_{rq}$$

And

$$Q_s \approx \frac{3}{2}|V_s|\left[\frac{|V_s|}{\omega_e L_s} - \frac{L_m}{L_s} i_{rd}\right]$$

Equations $P_s \approx -\frac{3}{2}\omega_e \Psi_s \frac{L_m}{L_s} i_{rq}$ and $Q_s \approx \frac{3}{2}|V_s|\left[\frac{|V_s|}{\omega_e L_s} - \frac{L_m}{L_s} i_{rd}\right]$ showed that the stator active power is controlled with the q component of the rotor current and the stator reactive power with the rotor current d component. In figure can be seen both hybrid fuzzy controllers to regulate the d and q rotor current components.The real platform to test the double feed induction generator and its main characteristics used also for the simulation purpose are shown in figure.The real system is based on a DS1103 board and is programmed using the software Matlab/Simulink. The board controls the inverters in a Back to Back configuration generating the SVPWM pulses. The grid connected inverter, is regulated keeping the DC bus voltage constant. The speed of the DFIG is measured with a 4096 impulse encoder connected to the DS1103 using the frequency method.

First, the inner current loops are adjusted. The used hybrid fuzzy controller is the same as have been used in the squirrel cage machine. The scaling factors have been adapted after realizing multiple simulations and finally adjusted in the DFIG test rig.

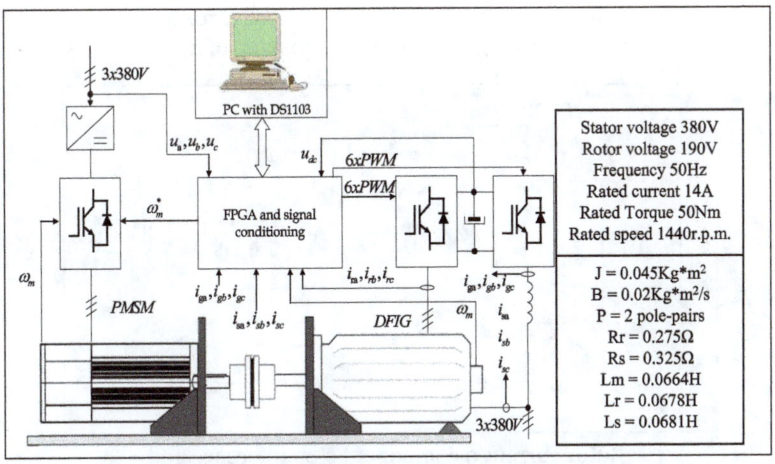

DFIG rig test and its main characteristics.

To test the performance of the hybrid fuzzy controller it will be compared to a conventional PI controller. In a classical PI controller the proportional term for a bandwidth of 3000 rad/s and a phase margin of 80° with the machine parameters given in figure is 0.015. For both current controllers, the proportional term KP will be 0.015 and the scaling factors are GE = 300, GCE = 0.025 and GCU = 0.2. The regulators maximum and minimum limits are ±1, equivalent to ±310 V per phase in the rotor.

Figure shows the hybrid fuzzy rotor q current controller surface and the trajectory when a step reference from 10 to 20 amperes is produced. The feedback or real rotor q current reaches the real value quickly, it takes around 3 ms. The trajectory on the fuzzy surface for this step shows how the error is moving around the high slope where the error is close to zero. In table, the performances of two controllers are summarized. The hybrid fuzzy and the conventional PI have similar dynamic response, showing the fuzzy controller a better performance when IAE, ITAE, ISE and ITSE indexes are used to evaluate the performance.

In a DFIG control there are two operating regions depending on the wind speed. Below the machine rated power, the blade pitch angle is set to zero degrees to get the maximum power. When the wind speed is sufficiently fast to get power from the wind higher than the rated power, enters into the second region. In this region the blade pitch angle controller regulates the output power modifying the pitch angle to get the rated power from the generator without damage it.

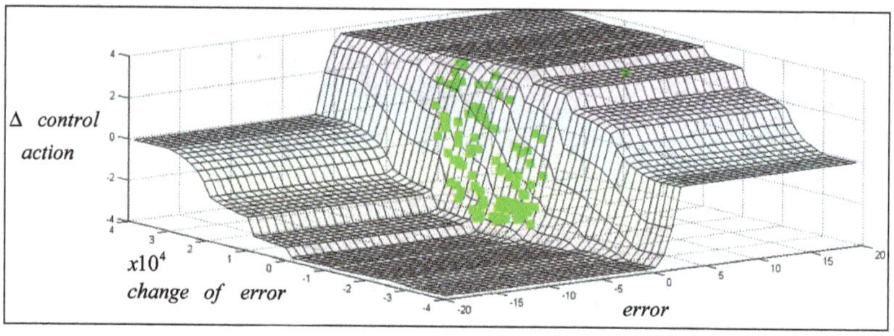

Rotor current q component controller fuzzy surface and trajectory after current step of figure.

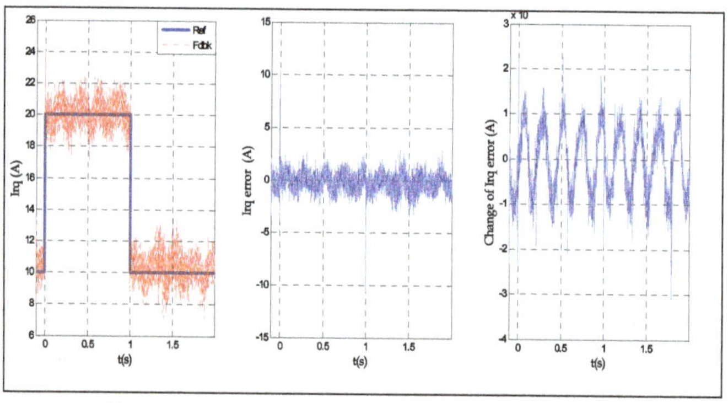

Rotor current q component step reference and feedback, error for the step, and change of error.

Table:Time domain specifications and performance criteria for two classes of controllers.

	Delay Time	Rise Time	Settling Time	% Overshoot	IAE	ITAE	ISE	ITSE
PI	1.2ms	1.2ms	4ms	20	8413	4110	11576	5105
FPI+P	1ms	1ms	3.5ms	20	7925	3880	10200	4510

The power transmitted to the hub of a wind turbine can be expressed as,

$$P_{turb} = \frac{1}{2} C_p(\lambda, \beta) \rho_{air} \pi R^2 v_w^{\ 3}$$

Where, ρ_{air}, is the mass density of the air, R is the radius of the propeller, Cp is the power performance coe4fficient, v_w is the wind speed, β is the pitch angle and is the blade tip speed ratio and is defined as,

$$\lambda = \frac{R \cdot \omega_{pr}}{v_w}$$

and ω_{pr} is the angular velocity of the propeller.

The power performance coefficient Cp, used according to the tip speed ratio and the pitch angle for the DFIG is shown in figure.

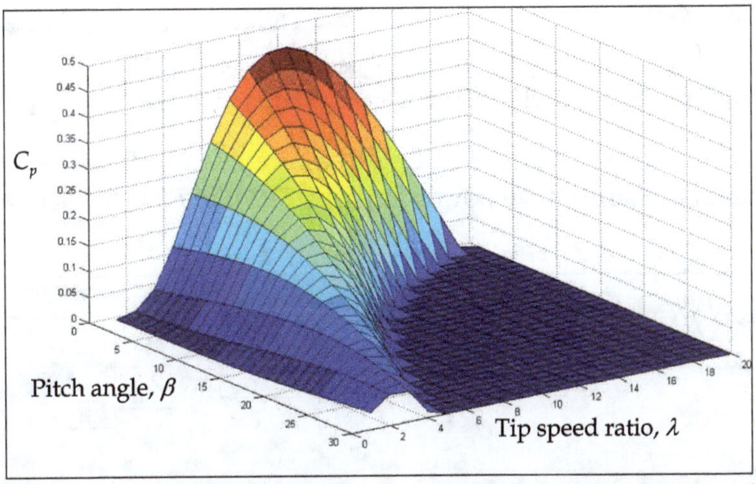

Power performance coefficient depending on tip speed ratio and pitch angle.

Figure shows for a pitch angle of 0° the obtained power from the wind according to the propeller speed. The black line indicates the maximum power and the propeller speed to get this power from every wind speed. When the obtained power reaches the machine rated power, the wind energy is wasted changing the pitch angle and getting the rated power.

For a known wind speed and using figure, the propeller optimum speed and the power are obtained. Then, with equation 30 the rotor q component is determined as reference.

The inertia of the blades turned by the drive is large and a real pitch actuator has thus limited capabilities. Its dynamics are non-linear with saturation limits on pitch angle (usually from 0 to 30°) and pitching speed rate around 10°/s.

The actuator is modelled in closed loop with saturation of the pitch angle and a pitch rate limitation. This closed loop configuration with integrator, gives similar result as a first order transfer function but with limitation of the pitch rate. If the pitch reference angle is outside the lower and higher limits, the integrator output is prevented from growing indefinitely.

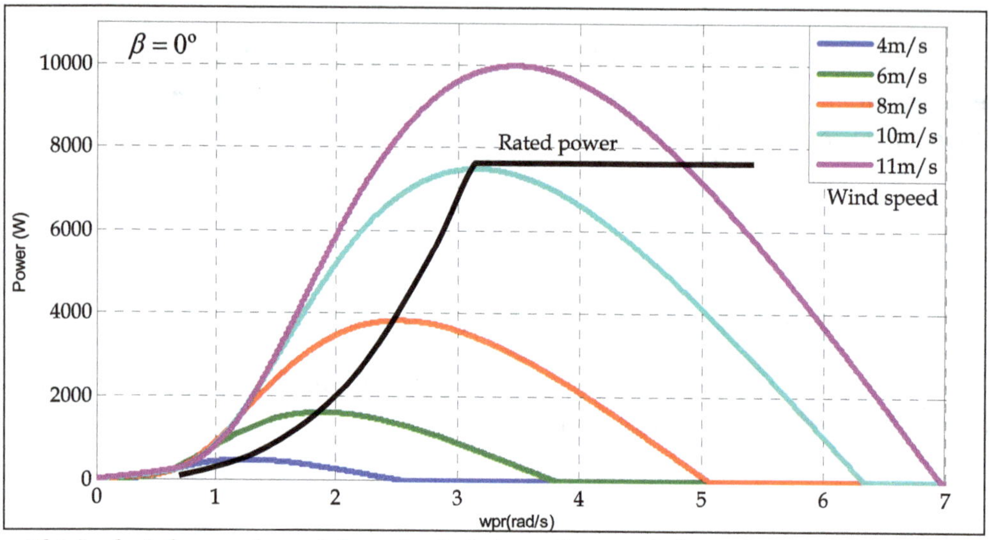

Obtained wind power for a pitch angle of 0°, depending on wind speed and propeller speed.

The pitch control diagram is shown in figure, where P is the DFIG real power, Pmax DFIG is the maximum admissible power for the DFIG and P* is the active power reference.

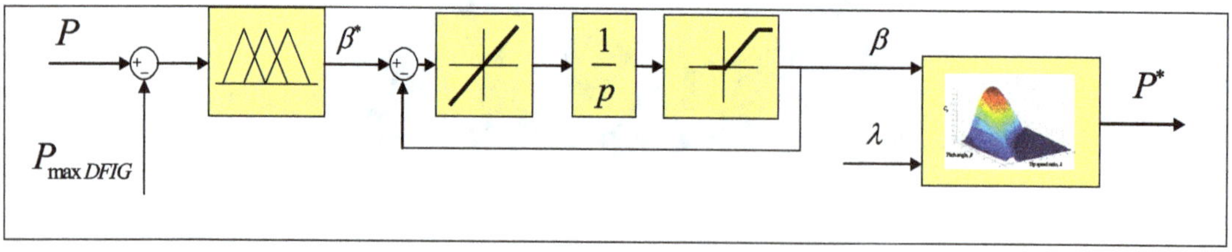

Pitch control diagram.

The pitching speed rate is fixed to 10°/s, the pitch angle is limited from 0 to 30°, the KP value and the scaling factors adjusted after simulations ensuring stability for the pitch controller are KP = 0.003 , GE = 400, GCE = 0.24 and GCU = 0.1 . The hybrid fuzzy regulator maximum and minimum limits are 0 to 30° as pitch angle reference limit.

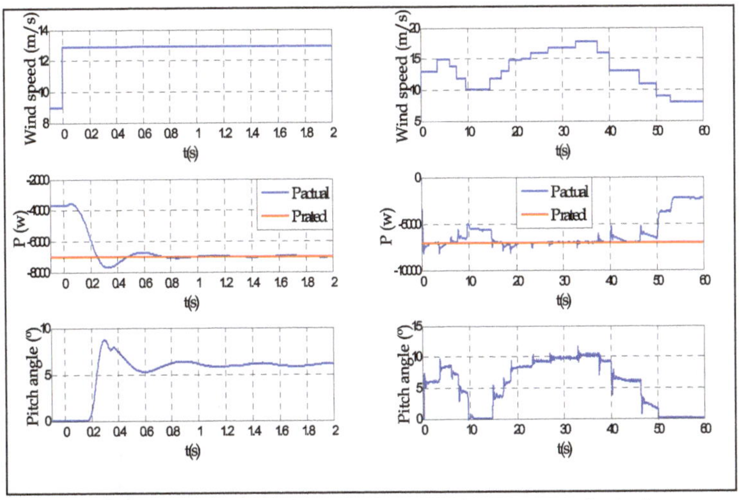

Fuzzy pitch control performance when a step of wind speed and a random wind speed profile are produced.

Figure left, shows the response of the pitch control when a wind speed step from 9m/s to 13m/s is produced. The obtained total power from the wind at 9m/s is 3800w and when the wind speed power is higher than the fixed 7000w, the pitch angle starts the regulation to limit the total power. The figure to the right shows the same signals for a random speed profile. When the wind speed is lower than 10m/s the pitch angle is zero, and all wind power is converted in electric power, but when the speed is higher, the pitch angle is regulated limiting the maximum power returned to the grid.

Parameter Varations

As it was commented into the introduction, the fuzzy logic approach is based on linguistic rules, and the controller robustness is high. To verify the above, the squirrel-cage motor is replaced by a different one. The motor parameters change and without realizing any adjustment in the controllers the speed regulation is tested in a motor control with conventional PI controllers and with the proposed hybrid fuzzy controllers. The new motor parameters are: Rr = 1.2 Ω, Rs = 1.5 Ω, Lm = 0.108 H, Lr = 0.12 H, Ls = 0.12 H, J = 0.038 Kg*m^2.

Figure shows the speed of the machine when there is a big noise in the stator alfa and beta components; in fact the noise is very high. The speed reference is 1000 rpm and a load step of 40 Nm is applied to the new machine, without readjusting the controllers, at 0.5s. The left figure shows the response of the machine controlled with PI controllers. The performance of the system becomes wrong when the load changes after 0.5s, the system becomes instable. Instead, in the right figure the motor is controlled with the hybrid fuzzy controllers adjusted. When the load torque is applied to the machine the speed regulation after that moment is correct. This is an example of the robustness of the fuzzy controller compared with the conventional PI controllers when there is noise in the measurements, in this case stator current measurement.

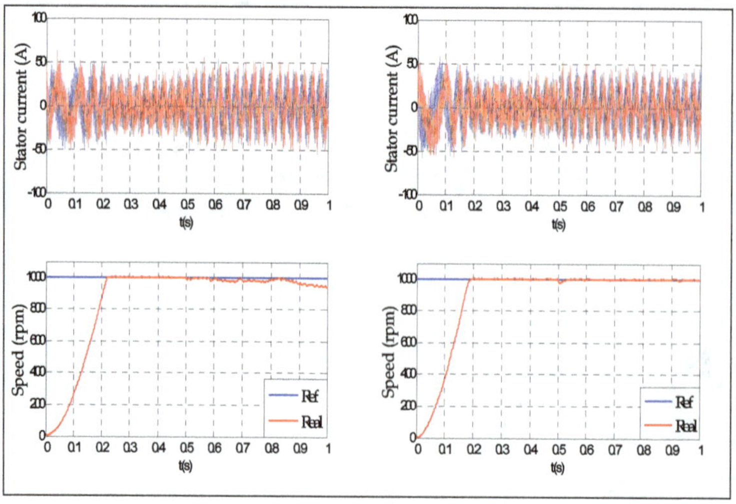

Speed regulation when there is a big noise in the stator current measurement. A load step of 40Nm is applied to the machine at 0.5s. Left, PI controllers. Right, hybrid fuzzy controllers.

References

- Fuzzy-locating-system: wiki.gis.com, Retrieved 09 June, 2019
- Intelligent-Neuro-Fuzzy-Application-in-Semi-Active-Suspension-System-224829069: researchgate.net, Retrieved 13 April, 2019
- Salmeron, Jose L. (2012). "Fuzzy Cognitive Maps for Artificial Emotions Forecasting". Applied Soft Computing. 12 (2): 3704–3710. doi:10.1016/j.asoc.2012.01.015
- Application-of-fuzzy-logic-in-control-of-electrical-machines, fuzzy-logic-controls-concepts-theories-and-applications: intechopen.com, Retrieved 28 May, 2019
- Adaptive fuzzy controller for a class of nonlinear systems, International Journal of Innovative Computing, Information and Control 1(2): 727–742
- Acampora, Giovanni; Loia, Vincenzo; Lee, Chang-Shing; Wang, Mei-Hui, eds. (2013). On the power of Fuzzy Markup Language. Vol.296. Studies in Fuzziness and Soft Computing. 296. Springer. doi:10.1007/978-3-642-35488-5. ISBN 978-3-642-35487-8
- Control-and-estimation-of-ac-induction-machines-using-fuzzy-logic, fuzzy-logic-controls-concepts-theories-and-applications: intechopen.com, Retrieved 23 July, 2019
- Gerla, Giangiacomo (2005). "Fuzzy logic programming and fuzzy control". Studia Logica. 79 (2): 231–254. CiteSeerX 10.1.1.103.1143. doi:10.1007/s11225-005-2977-0

PERMISSIONS

All chapters in this book are published with permission under the Creative Commons Attribution Share Alike License or equivalent. Every chapter published in this book has been scrutinized by our experts. Their significance has been extensively debated. The topics covered herein carry significant information for a comprehensive understanding. They may even be implemented as practical applications or may be referred to as a beginning point for further studies.

We would like to thank the editorial team for lending their expertise to make the book truly unique. They have played a crucial role in the development of this book. Without their invaluable contributions this book wouldn't have been possible. They have made vital efforts to compile up to date information on the varied aspects of this subject to make this book a valuable addition to the collection of many professionals and students.

This book was conceptualized with the vision of imparting up-to-date and integrated information in this field. To ensure the same, a matchless editorial board was set up. Every individual on the board went through rigorous rounds of assessment to prove their worth. After which they invested a large part of their time researching and compiling the most relevant data for our readers.

The editorial board has been involved in producing this book since its inception. They have spent rigorous hours researching and exploring the diverse topics which have resulted in the successful publishing of this book. They have passed on their knowledge of decades through this book. To expedite this challenging task, the publisher supported the team at every step. A small team of assistant editors was also appointed to further simplify the editing procedure and attain best results for the readers.

Apart from the editorial board, the designing team has also invested a significant amount of their time in understanding the subject and creating the most relevant covers. They scrutinized every image to scout for the most suitable representation of the subject and create an appropriate cover for the book.

The publishing team has been an ardent support to the editorial, designing and production team. Their endless efforts to recruit the best for this project, has resulted in the accomplishment of this book. They are a veteran in the field of academics and their pool of knowledge is as vast as their experience in printing. Their expertise and guidance has proved useful at every step. Their uncompromising quality standards have made this book an exceptional effort. Their encouragement from time to time has been an inspiration for everyone.

The publisher and the editorial board hope that this book will prove to be a valuable piece of knowledge for students, practitioners and scholars across the globe.

INDEX

www.ingramcontent.com/pod-product-compliance
Lightning Source LLC
LaVergne TN
LVHW080617210926
848485LV00003B/171